Give Us Liberty

Give Us Liberty

A Tea Party Manifesto

Dick Armey
and Matt Kibbe

An Imprint of HarperCollinsPublishers

HarperCollins books may be purchased for educational, business, or sales promotional use. For information please write: Special Markets Department, HarperCollins Publishers, 10 East 53rd Street, New York, NY 10022.

FIRST HARPERLUXE EDITION

HarperLuxe™ is a trademark of HarperCollins Publishers

Library of Congress Cataloging-in-Publication Data is available upon request.

ISBN: 978-0-06-201817-5

10 11 12 13 14   ID/RRD   10 9 8 7 6 5 4 3 2 1

To the sons and daughters of liberty
who did not die with Sam Adams but are active
and organizing across the country to
protect our freedoms to this very day.

Contents

Contents

Give Us Liberty

PROLOGUE: REVOLUTION AND RETREAT

Now, we Americans understand freedom. We have earned it, we have lived for it, and we have died for it. This nation and its people are freedom's model in a searching world. We can be freedom's missionaries in a doubting world. But, ladies and gentlemen, first we must renew freedom's mission in our own hearts and in our own homes.

—BARRY GOLDWATER

The pendulum of politics is always swinging. Republicans gain seats, Democrats win them back, and the struggle continues. The one thing that seems inevitable no matter the party in office is the continued growth of government. Occasionally, however, a fresh jolt of conservative energy interrupts that growth.

I have witnessed three conservative revolutions in my lifetime, each of which taught important lessons.

Barry Goldwater came first, offering a choice, not an echo. Ronald Reagan was next and proved that small-government conservatism was the best path to peace and prosperity. Under Reagan, we regained our national confidence and our economy flourished. A few years later came the third wave, the 1994 Republican Revolution in Congress and its Contract with America that resulted in the impossible—a balanced budget.

I was a student in Oklahoma, studying economics, when Barry Goldwater ran for president in 1964. Never had I heard a political figure speak so passionately on freedom, the genius of the Founding Fathers, and our enduring values as a nation.

That same summer I heard Ronald Reagan's classic campaign speech, "A Time for Choosing." During this time I was so engrossed in my studies, I almost never thought of politics. But Goldwater was different; and for the first time in my life I was excited and inspired to participate in the political process.

I did not think of myself as a conservative. I thought my belief in personal and economic freedom, respect for the Constitution, gratitude for those who served in the armed forces, and trust in my fellow citizen was just common sense. My beliefs and values were shaped by my upbringing. I was born in Cando, North Dakota, the fifth of eight children. My parents operated a rural grain elevator, and this is where I developed my interest in economics. I remember my parents quieting all the children down at one o'clock every day so they could hear the grain market report coming in. "Oats up, wheat down, corn holding firm." This was the market, supply and demand, at work.

It was also during this time I remember post–World War II refugees from socialist Eastern Europe fleeing to the prairie. It struck me that people would leave their homeland and everything they knew to go someplace for freedom. I did not see anyone fleeing the United States to go in search of collectivism.

My father was an avid fisherman, and I would often join him for trips up north to Canada to fish for northern pike. The drive to our fishing camp passed through countryside dotted with painted barns straight from a Norman Rockwell canvas. But as soon as you crossed the border into Canada, I noticed the barns were unpainted. I wondered why Canadian farmers would allow their barns to degrade from exposure to the elements. The answer, I discovered, was government. At the time, Canada taxed painted buildings, so farmers left their structures exposed to avoid the penalty. These things make quite an impression on a child.

After a few years working as a lineman for a local utility, I decided to become the first member of my family to go to college. But I was stunned when the commonsense values I grew up with were militantly rejected by many on campus. It was during the Goldwater campaign that I learned of an elite who existed in government offices and college faculty lounges and who were hostile to the universal values of the American people. I could not believe that here in America a group of people believed they were entitled to redistribute wealth to satisfy their notions of social justice, regulate others' lives, and understand my best interests better than I.

Years later, when I was the chairman of the economics department at North Texas State University (now the University of North Texas), I sat through a faculty meeting in which a professor claimed part of our job was to teach students how to emote. I was stunned. I simply wanted to teach economics and did not see why feelings should be taught in my classroom. It was time for me to find another line of work.

One beautiful Texas evening in 1984, I found myself watching C-SPAN. A procession of representatives were speaking from the House floor in opposition to the president's fiscal policies. Their arguments were dangerously incorrect—they would doom our country to a collectivist quagmire. I figured Ronald Reagan could use reinforcements, so I decided to take a sabbatical year and run for Congress. In what is still considered a huge upset, I defeated the incumbent and went to work in Washington.

As time passed, each boom gave way to a bust. The iron triangle of entrenched politicians, bureaucrats, and the motivated network of special interests was temporarily slowed but never defeated. It seems inevitable that as the state expands, freedom is eroded. Government growth continues, and each new effort to check its expansion faces a larger and larger challenge.

I believe we are at a turning point in our nation's history, as the pendulum has swung far to the left. Trillion-dollar deficits, government control of health care, federal ownership of banks and auto companies, taxpayer-funded bailouts of irresponsible home owners, and attempts to control energy consumption have combined to push a nation conceived in liberty and devoted to free markets to bankruptcy, down what F. A. Hayek called the road to serfdom.

But I believe Americans are genetically opposed to big government. They won't accept it, and they have been joining with their fellow citizens in the streets to take America back. I believe this movement, the Tea Party movement, has the opportunity to break the boom-and-bust cycle and restore a constitutionally limited government and bring fiscal sanity to Washington.

Republicans in the 1980s could not even dream of setting the legislative agenda. They had not been in a position of leadership since 1954 and had become complacent. Senior members of the party were satisfied if a Democratic chairman occasionally left them a few crumbs. Life was comfortable in the minority as long as you did not rock the boat. No one was accused of partisanship because the majority always got its way. Members received their perks, such as travel abroad and special banking privileges, and based their political careers on parochial pork projects.

Things began to change in the early 1990s, when a rebellious group of small-government backbenchers began a hostile takeover from within the Republican House caucus. We made life difficult for the establishment old bulls in the party because we thought they were too complacent. The Republican leadership was always having to apologize to the Democrats for us.

We believed it was time to bring restraint to Congress, and we set a goal of retaking the House. Newt Gingrich and I believed we would be successful in the 1994 elections if we were able to prove to the American people that we had a national policy vision. My office set about to draft a contract based on policies that were important to the American people and were blocked by the Democrat majority. Newt thought we should focus on ten items.

The Contract with America, as it later came to be known, outlined our platform of limited government. The contract nationalized the vision of the Republican Party in a way that unified our base and appealed to independents. We championed national issues that were good for all Americans, not just special interests. This vision was validated when Republicans took control of both houses of Congress.

In our first press conference after we were sworn in, I was asked what I was going to do with all the power the American people had just given us. I simply responded that the American people did not give us power, they gave us responsibility.

And for a few years, we did pretty well.

As in any enterprise, the spirit of 1994 eventually fell victim to the natural life cycle of a firm: the passion and creativity of the entrepreneurs give way to the complacency of the bureaucrats. The Republican vision for America became a parochial vision about the careers of the individual members of the majority party. By the summer of 1997, the appropriators—rightly called the "third party" of Congress—had begun to pass spending bills with Democrats. As soon as politics superseded policy and principle, the avalanche of earmarks that would eventually crush the majority began.

Balancing the budget, eliminating unneeded and/or unconstitutional programs, and limiting the overall growth of government took a backseat to an explicit strategy to use tax dollar appropriations to trade votes on the House and Senate floors and to use politically motivated earmarks to buy reelection back home. It became "all about me" (the politician), my needs (comfort, power, job security) and my goals (government power bought with other peoples' money).

The late House Speaker Tip O'Neill once famously quipped "all politics is local." That is great advice for liberals, whose base is looking for special favors and handouts. But the constituency that supported the contract, conservatives of both parties and independent voters, were not interested in local pork, but rather were looking for good policies that benefited the entire nation.

At a leadership meeting in early 2002 I pulled out my yellow schedule card and scribbled a note to myself: "Every week we come to town and do things we ought not do in order to keep the majority so we can do the things we ought to do but never get around to doing." It was time to hang up my congressional spurs, so I soon announced my retirement from Congress. But I wasn't going to give up my life's work in liberty. Citizens for Sound Economy (now FreedomWorks) had been with me in every important fight defending economic freedom, so it made sense for us to join forces.

While temporarily effective, the Republican Revolution of 1994 devolved into an embarrassing gap between Republican rhetoric and fiscal reality.

Looking back, the revolution of 1994 was an insider takeover. Inside jobs like 1994 are an inherently weak strategy because they are too dependent on the good intentions of people astride the levers of power. It is best described as the benevolent despot model of social change. "If we just elect the right people" is an appealing myth. The inevitable unraveling as leaders become compromised leaves the nation exposed to the boom-and-bust cycle of top-down legislative change. The "great man" effect leads to inevitable disappointment as elected leaders fall victim to the incentives that come with holding office.

The nature of big political turnover and citizen revolt has always been inherently reactionary. At some point the veil is lifted and people notice, en masse, some gross mismanagement of the public trust and the public's purse. Scandals, arrogance, budgetary abuse, and economically destructive policies bring citizens out of their homes and to the polls. They throw the bums out. Yet in many cases, the damage is already done and new programs are in place, never to be repealed. The constitutional limits of the state are reduced even further. Throwing the bums out this way will produce a new generation of bums that will face the same incentives to disappoint expectations and abuse the public trust.

But the current groundswell of small-government activism, commonly referred to as the Tea Party movement, has the potential to permanently change this paradigm. We all know we face a fiscal time bomb squarely that is threatening the American way of life. The government has grown too big for the private sector and citizenry to sustain it. We also know we cannot go back to the boom-and-bust cycle but instead must change politics.

The Tea Party movement has the power to break the cycle by establishing a constituency standing at the dead center of American politics. This is a true bottom-up revolution. It does not need formal leaders or a hierarchy; all it needs is sound limited government principles and a dash of practical American intuition. Politicians will inevitably disappoint you, but ideas will not.

When Tea Partiers show up, they do not demand any special favors and more government but literally show up with signs that read WE WANT LESS! And it's not a moment too soon.

. . .

I wrote this book for my grandkids. When I was their age, my grandparents looked at my bothers and sisters and me and were filled with hope for our futures. We dreamed big, full of rugged American individualism and a can-do spirit.

I want to pass on a strong and confident nation to my grandchildren. But I fear a nation crippled by debt. When they start families of their own, how are they going to take care of themselves and also repay the trillions we're spending right now? What will they think of us for borrowing half of the federal dollars we spend?

I am convinced that this generation of patriot activists is ready to defend the values that made our country great. Just ask anyone who marched down Pennsylvania Avenue on September 12, 2009, as many of the people reading this book did. That day, nearly a million Americans gathered after organizing their communities and traveling at their own cost to make a statement, together, that enough was enough. I marched with folks from all over the nation, all backgrounds and ethnicities—grandmothers and college students, small-business owners and soldiers, Republicans, Democrats, independents, libertarians, and evangelicals—all marching hand in hand. It was America at its best. Our neighbors were standing together to defend personal liberty and economic freedom.

We are a new and permanent force in American politics. And we want our liberty back. But the powerful in Washington won't relinquish control easily. We have to take it. And that is going to mean a lot of work. My hope is that this book will empower the millions of Americans who want to make a difference to do so.

In these pages you'll find inspiring stories about citizen activists who have dedicated themselves to doing just that. You'll learn what inspired them. You'll read firsthand accounts of the actions people take from across the country who decided enough is enough. You'll see how the intellectual underpinnings of our limited government movement have time and again led us to the right positions while "experts" inside the Beltway said we were wrong. You'll understand that our belief in limited government puts us at the center of American politics—not on the fringe—meaning we will be the one who decides who wins elections. And you will also read our FreedomWorks Grassroots Action Toolkit that will make you an effective and powerful patriot-activist.

Our Founding Fathers entrusted us with the preservation of our liberty. Together, we can save the republic they created.

DICK ARMEY
Dallas, Texas

1

THE SILENT MAJORITY SPEAKS

I was so excited we were really doing it. We got to the Harborside Center and all we could see was security and people lined up to see the speech; we were the only protesters. I told Ron that I was right—we'd be the only ones there that day. But we stuck to the plan and he dropped me off at the front entrance and left to park the car. Within minutes, a Fort Myers police officer got right in front of me and told me I had to move. I was the only one there! Can you imagine anyone believing that a woman like me was a threat? I knew right then and there we were doing the right thing. Before long we had six or seven people with us. The media found out about us and I ended up talking to a producer from Fox News. I guess the rest is history.

—MARY RAKOVICH

It's difficult to say exactly when the modern Tea Party movement came into being. In a way, the question itself is incorrect. It's often asked by the same observers who labeled the movement a "Tea Party," the ones who ignored it for months until they could no longer hide from the truth. They would like to identify an exact beginning. That would make it easier to track the middle and the end, the conclusion to a phenomenon they believe to be a fluke, a flash in the pan, a temporary surge from the lunatic fringe. It's none of those things. We believe that Americans have always stood up for liberty and possess an innate sense of responsibility to guard their freedoms. In other words, we've always been here. And we're not going away.

AN UNLIKELY HERO

Before her groundbreaking Florida protest in February 2009, Mary Rakovich was better known as a caregiver and animal lover. A fifty-three-year-old grandmother, Mary had worked as an automotive electrical engineer in Michigan before she was laid off in 2005. Unable to find work after several interviews and concerned for ailing in-laws, she and her husband, Ron, relocated to Florida to care for her family.

For most of her life, Mary had remained too busy for politics. Idle discussion about current events was a luxury reserved for those who weren't working to support a family and care for aging parents. What she did feel passionate about was her family and the commonsense principles that guided them. In recent years, however, she noticed that her government did not share her values—in fact, it appeared to be heading in the wrong direction. Enamored with elaborate entitlement programs and endless pork barrel projects, legislators in Washington seemed interested in anything but balancing a budget.

By March 2008 she was hearing staggering numbers on the news. The federal government had ballooned under President Bush. Mary had always held concerns about deficits and spending, but for the first time, it started to really concern her. "How would we pay all of this money back?" she wondered.

As she watched from her living room in Cape Coral, the subprime housing boom slowed, staggered, and crashed. Hoping to find her elected leaders prepared to make hard choices, she was instead disappointed with their reactions.

"President Bush listened to his advisors and made mistakes," Mary said. "The bank bailout was ridiculous. If you can't pay your bills and your business model has failed, you simply close your doors. That was the way it was supposed to work in our system."

Mary spent increasing time researching government and economic policy. With the presidential election approaching, she was looking for leaders but was disappointed by the major-party candidates. Senator Barack Obama supported the Bush bailouts. She watched in disbelief as Senator John McCain "suspended" his campaign at the end of September to head back to Washington, not to restore fiscal order but to join in and make sure the bailouts passed. She had volunteered for his presidential campaign but felt he took the wrong position on the financial crisis.

Mary was starting to feel frustrated and angry. "The principles that this country were founded upon were being eroded—freedom was being eroded—and I needed to do something about it," she said.

After the election, Mary continued her research. She reread the Constitution and brushed up on American history. "I realized how much I believe in the founding principles, how much they mean, how precious our freedom really is."

Fresh off the campaign trail, President Obama began to push for a trillion-dollar stimulus bill. For Mary, this was simply too much. It seemed the new president was taking a risky bet with other people's money, recklessly hoping for a different result than the various Bush administration stimulus packages and bailouts had already delivered. She was upset and concerned for her children and grandchildren. Inspired to act, Mary began searching for more ways to make her voice heard.

She soon discovered and joined FreedomWorks, the grassroots group we lead, and learned that an activist training seminar was scheduled to take place in Tampa in the coming days. The Rakoviches signed up and attended the workshop with about eighty other Florida citizens. The three-hour training session covered grassroots basics, from calling local talk shows to recruiting activists, hosting an event to using online tools like Twitter and Facebook to build a community of activists.

"The training gave me hope, and I left energized. I saw that others were as concerned as I was. I could see that I was not alone."

The Rakoviches were talking about putting their new training to use when our colleague Brendan Steinhauser of FreedomWorks called to follow up and support their efforts. He urged them to take it to the streets, saying, "You only need the two of you and a few signs to make your voices heard."

It did not take long for an opportunity to use their newly acquired skills to present itself. Less than two weeks after the FreedomWorks training seminar, campaign manager Nan Swift learned that President Obama was scheduled to appear in Fort Myers to extol the virtues of his stimulus plan. He would be joined by Charlie Crist, Florida's Republican governor, who had enthusiastically supported the proposed additional federal deficit spending.

They targeted the Obama/Crist rally and got to work. Ron called a local political talk show to announce the protest while Mary reached out to other activists to join them. They scouted the event location, noting the arrangement of barricades and likely restrictions from police and Secret Service personnel. They called local police to make sure a protest would be allowed and to find out about any special restrictions. The night before the rally, Mary gathered poster boards and markers to make signs. Even then, she still had misgivings.

"I thought, What am I doing here? I'm not a protester. Are we going to be the only people out there?"

Mary remembered the speaker's advice at the FreedomWorks seminar: *Have fun with it.* She opened the box of markers and wrote REAL JOBS, NOT PORK in big black letters across a poster board. It felt great to *do* something! Come hell or high water, she would be there to greet the president. On February 10, 2009, Mary and Ron Rakovich set out for the Harborside Center in Fort

Myers with signs, a cooler of water, and the courage of their convictions.

It was a beautiful Florida day, with clear skies and a light breeze off the river. Attendees were lining up at the front entrance and police could be seen directing traffic. Sticking to their plan, Ron would drop Mary off and park the car while she set up signs and sought out fellow protesters. But as soon as Ron pulled away, a Fort Myers police officer instructed her to move.

"From the beginning, they did not want us engaging the attendees," Mary explained. "I was told to move to the back parking lot, but the stimulus supporters could stay." Surprised and angry at being told to move three separate times, she was moved to tears. "I have every right to be here," she exclaimed to the officers. "I'm just one woman with a sign. Why can't I voice my opinion?"

Undaunted, she moved from the entrance and adjusted the volume of her commentary accordingly. One by one, other protesters began to join her and put her signs to good use. Surrounded by Obama supporters, the group stood their ground and did what they came to do.

Nearby reporters were soon intrigued by the protest. It was obvious at first glance that this was not one of the usual mobs that descend on presidential events. No one

was screaming obscenities or proclaiming the usual opposition to war, oil, red meat, or all of the above. These
were middle-class Americans of all ages talking about
fiscal sanity. And they were *making sense.*

Mary was contacted by a producer at Fox News
shortly after the event. She was featured on a newscast
later that day talking about the protest. It was her first
time in front of a microphone, let alone on live national
television, and now she was very nervous.

"I was probably shaking," she recalled. "The crew
asked me to take my glasses off for the remote feed. I
couldn't see the camera and was being asked questions
through a small earpiece. I could barely hear him. The
last time I think I spoke in public was at a preschool
parent-teacher meeting."

The effect was remarkable. Mary's honest sincer-
ity and obvious lack of preparation charmed and intrigued viewers. They listened to what she said and
realized they agreed. To the millions who were just
as outraged as Mary but unsure of how to make a
difference, a role model had been found.

As it turned out, Mary was far from alone. Years of
broken budgets and wasteful spending had created vast
reservoirs of discontent. Conventional wisdom holds
that the government's fiscal irresponsibility, deficits,
and federal spending are perennial issues in American

politics that never move large numbers of people off the couch or voters to the polls. That assumption was about to be put to the test.

ECHOES ACROSS THE COUNTRY

More than three thousand miles away, Keli Carender was frustrated. A Seattle schoolteacher and member of a local comedy improv troupe, Keli had always been interested in politics and current events. But she had never seen herself as an activist, leaving the marching to others while she aired her views among friends in homes and coffee shops. The closest she had ever come to political demonstration was a weekend back in high school when she participated in a Washington Girls State convention hosted by the American Legion. By spring 2009 Keli was reconsidering her responsibility as a citizen.

"Our nation's fiscal path is just not sustainable," she said. "You can't continue to spend money you don't have indefinitely."

She watched what passed for debate over the stimulus package on C-SPAN and reached for the phone, determined to express her objections to her representative. But she rarely got through, and when she did, the congressional staffers were condescending and sarcastic. They acted as though she couldn't possibly understand the workings of the national economy and rather should leave such matters to the professionals.

Keli decided it was time to make a choice. "I could give up and be depressed watching my county commit fiscal suicide, or I could find a way to speak up." She chose the latter. "I figured that if the Left could use protests to get their message out, so could I."

Demonstrating the blunt practicality that would typify early activist events, Keli called the local police department in February 2009 and simply asked, "How do you do a protest?" The parks department graciously walked her through the request and permitting process. In just five days she set up a "Porkulus Protest" in downtown Seattle. The next several days were spent reaching out to anyone and everyone who might be willing to publicize or participate. "I called think tanks, clubs, and radio hosts, anyone I could think of that may be interested in coming out," Keli recalled. She caught a break when popular conservative commentator Michelle Malkin promoted the event on her blog.

On the morning of the event, Keli was nervous that only her parents would show up. But thanks to her work, 120 citizen activists, many of them protesting for the first time, took to the streets to draw attention to fiscal irresponsibility in Washington. "We brought barbecue pork sandwiches," she recalled with a laugh. "And they did not go to waste."

At FreedomWorks' headquarters in Washington, D.C., e-mails and phone calls began pouring in. From Tampa to Seattle, people were publicly demanding accountability from their elected leaders. To those who were paying attention, a clear theme had emerged: Mary, Keli, and their fellow citizens across the country live on budgets. They don't spend more than they earn. When times are tough, they make do with less. They expect their government to do the same.

ENOUGH IS ENOUGH

At the turn of the millennium the U.S. national debt stood at $5.6 trillion. By the end of 2008 that amount had nearly doubled to $10 trillion, which translates to more than $85,000 per household. By 2018 the deficit is projected to nearly double again to more than $18 trillion.

In 2000 the cumulative deficit was just below 60 percent of Gross Domestic Product (GDP). According to the Congressional Budget Office, that percentage is expected to rise to more than 100 percent by 2012. By comparison, France's 2010 debt-to-GDP ratio is 84 percent.

Since 2003 the total national debt has increased by more than $500 billion each year with shocking increases of $1 trillion in 2008 and $1.9 trillion in 2009. Not surprisingly, this spending is far outpacing the growth of the economy. In 2000 the federal budget was $1.6 trillion. By 2009 the budget had expanded to $3.6 trillion but only took in $2.1 trillion in revenue. This means that the federal government is now borrowing nearly fifty cents of every dollar it spends.

There are only three ways for the government to spend money it does not have: it can raise more tax revenue; it can borrow money from the private sector or from other nations; or it can debase the dollar by printing more currency. Make no mistake: Americans are aware of the crushing debt burden we are amassing as a nation. Individuals and families know that you can't spend your way to prosperity with money you don't have. They know this because they live it in their daily lives. The national debt threatens our future with crippling new taxes, sky-high interest rates, out-of-control inflation, and an increasingly weak dollar. This process punishes workers, savers, and investors alike. Ultimately, the world could lose its faith in U.S. currency, essentially bankrupting a great nation because our out-of-touch political class lacked the will to set priorities and live within a budget.

Government spending has been a concern as long as there has been a federal government. But recent events have elevated what was once an ongoing concern to the level of historic crisis. The response from the grass roots? *Enough is enough.*

THE RANT HEARD 'ROUND THE WORLD

In the wake of Mary's and Keli's events in February 2009, a movement was reawakened. Confused by the commonsense rhetoric and nonviolent, law-abiding tactics, political pundits and media observers were at a loss. The phenomenon needed a name.

An on-air commentator for cable news network CNBC, Rick Santelli was a fixture at the Chicago Mercantile Exchange, where he offered news and commentary on corn futures, yield rates, and other market data. On the morning of February 19, 2009, news coverage was dominated by Obama's proposal for yet another housing bailout. CNBC studio analysts calmly reported the news, discussing vast sums of taxpayer money in a tone ordinarily reserved for reporting on weather patterns over the Midwest. Standing by for a floor report, Santelli heard the commentary on his earpiece and began to fume.

After reporting on the latest housing bailouts, an anchor tossed to Santelli for his usual update. Santelli unexpectedly unleashed an impassioned rant.

"The government is promoting bad behavior!" Santelli shouted. "This is America! How many people want to pay for your neighbor's mortgages that have an extra bathroom and can't pay their bills? Raise your hand! President Obama, are you listening? You know Cuba used to have mansions and a relatively decent economy. They moved from the individual to the collective. Now they're driving '54 Chevys. It's time for another Tea Party. What we are doing in this country will make Thomas Jefferson and Benjamin Franklin roll over in their graves. We're thinking of having a Chicago Tea Party in July, all you capitalists. I'm organizing."

As he spoke, a group of traders formed around him on the trading floor. Capitalists to a man, they cheered the outburst and drowned out the planned transition, extending the segment and creating an indelible TV moment. Within hours, Santelli's rant had gone viral, earning more than a million views on YouTube and countless watercooler and dinner-table discussions across the country. The frustration that had been building, and which had begun to turn into street action, now had a name. The Tea Party was ready for the national stage.

CONCEIVED IN LIBERTY

Across the nation, private citizens who had never protested, never agitated, never taken a public political stand were gathering and organizing to make a difference. United by common principles and outraged by the complacence and indifference of their elected leaders, these individuals were ready to do something. Early meetings were filled with entrepreneurs, retirees, schoolteachers, civil rights leaders, lawyers, those who had prospered in recent years, and some who had fallen on hard times. All believed that the time to act had come, that their children and grandchildren deserved better and it was up to them to change the course of a nation.

But for all the excitement, the first wave of Tea Party activists faced significant challenges. They were poorly funded. They lacked national organization. They were greeted with skepticism by the political establishment. They included none of the political intelligentsia in their ranks, none of the gatekeepers and message experts and focus group gurus. How could they hope to influence a Congress of incumbent leaders with strong ties to interest groups and well-funded corporate backers? How could they challenge an administration that had swept into the White House with a landslide victory in the presidential campaign?

To many, the answer could be found in another group of unlikely activists who were overmatched and outgunned but fought anyway. Also comprised of ordinary citizens, this group had toppled an entrenched regime that seemed invincible. In fact, it had happened in 1773, right here in America.

2

THE AMERICAN REVOLUTIONARY MODEL

It was now evening, and I immediately dressed myself in the costume of an Indian, equipped with a small hatchet, which I and my associates denominated the tomahawk, with which, and a club, after having painted my face and hands with coal dust in the shop of a blacksmith, I repaired to Griffin's wharf, where the ships lay that contained the tea. When I first appeared in the street, after being thus disguised, I fell in with many who were dressed, equipped and painted as I was, and who fell in with me, and marched in order to the place of our destination.

—GEORGE HEWES, 1773

On a cold December morning in 1773 in Boston, Massachusetts, a group of concerned citizens gathered at the Old South Meeting House. Among their number was a poor shoemaker named George Hewes, a man of little standing in the bustling city but one who felt that his freedom was just as valuable as that of the wealthy merchants and landowners who were also present. At just five foot one, the diminutive Hewes had been denied military service due to his stature and had reluctantly settled into a trade he disliked. At heart, the man was an agitator. A veteran of the Boston Massacre, he was a devotee of Samuel Adams and the Sons of Liberty. He was no one special, just a man who felt his liberty was worth fighting for.

Infuriated by new taxes on imported goods and wary of the British troops stationed around the city, the men gathered that morning intending to take action. Too long ignored by Parliament and local officials seeking to protect their own positions rather than represent the citizens they were sent to govern, Hewes and his compatriots came to discuss a significant choice. Would they challenge British authority and stand up for what they believed was right?

After a heated debate, the meeting resolved that certain ships carrying vast quantities of tea should leave the harbor without the payment of any duty. The act would clearly signal their displeasure with the tax while falling short of any treasonous act that could result in fines, prison, or even hanging. Satisfied with their choice, the group sent a contingent of concerned citizens to report the message to the Customs House and force the release of the ships from the harbor. The collector of customs refused to allow the ships to leave without payment. When word of this decision reached the Old South Meeting House, a howl erupted from the hall.

Their bluff had been called. At this point, a compromise would be equal to surrender. What was planned as a peaceful expression of disagreement would now give way to protest. No permission would be sought, and the consequences would be accepted by every man in the room.

By early evening, a group of about two hundred men, some disguised as Indians, assembled on a hill overlooking the harbor. Bellowing war chants, the men marched two by two to the wharf, descended upon the three ships, and dumped their offending cargoes of tea into the water. Hewes gleefully pitched enormous crates of tea overboard, any one of which was likely worth more than a year of his income from shoemaking.

The reaction in London was swift and extreme. In March 1774, Parliament passed what later came to be known as the Intolerable Acts, which among other measures closed the port of Boston. The response at home was equally intense, as the Tea Partiers discovered a groundswell of public opinion in their support. Thousands of citizens had detested the arrogance and shortsightedness of British policy, but until then had not found a voice to speak out. Encouraged by the protesters' bold actions, public opinion galvanized against the Crown and in favor of separation.

In many ways, the American Revolution did not begin with a shot. It first echoed with a splash as crates of tea tipped into the murky waters of Boston Harbor. While the history books may remember the great leaders of the time and commemorate their achievements, a great deal of the glory must be shared with men like George Hewes—ordinary men who took extraordinary actions in defense of liberty.

A VOICE IN THE WILDERNESS

More than two hundred years later, Mary Rakovich and her fellow Tea Party activists were motivated by a similar sense of outrage. In fact, history teaches us that nothing could be more American than a protest. What made the opening salvos of the Tea Party movement so jarring to political, academic, and news media observers was its unlikely source—an irate group of citizens from across the political spectrum who were agitating and demanding change. For generations, guerrilla tactics had been a trademark of the Left, best demonstrated in ecoterrorism and virulent antiwar campaigns. Accustomed to Code Pink public disturbance stunts and blood-tossing animal rights' activists, students of political activism had come to understand public protest as left-leaning by definition and reserved for those who were willing to damage public property and disrupt legal activities. Now that middle-class Americans of all backgrounds were taking to the Internet, airwaves, and streets, conventional wisdom was turned on its head and the original Tea Party was seen in a new light.

For the staff at FreedomWorks, this was nothing new. Originally founded as Citizens for a Sound Economy (CSE) in 1984, the organization was dedicated to the idea that ordinary Americans could and should demand more accountability from their elected officials.

Long before Barack Obama employed community organizing tactics in the Democratic presidential primary states, our organization was mobilizing citizens in the fight for lower taxes, less government, and more freedom. In the 1980s we helped pioneer direct-mail techniques to engage citizens in direct action against big government. In the early 1990s we were one of the first issue advocacy groups to effectively use professional message development and paid television advertising to drive policy outcomes. We used these tools to help stop Vice President Al Gore's new tax on carbon-based energy, the so-called BTU tax.

As the years passed, we shifted tactics again to focus on organizing "boots on the ground," with paid field operatives, putting real people in front of political decision-makers. This proved instrumental in stopping the Clinton administration's attempted health care takeover. CSE organized protesters at every stop of First Lady Hillary Clinton's "Health Care Express," a national bus tour designed to galvanize support for the legislation. Instead of the adoring crowds she expected, Mrs. Clinton was greeted by hundreds of citizens opposed to her plan. We even followed her entourage with a truck towing an old broken-down bus. On the side was spray-painted THIS IS GOVERNMENT-RUN HEALTH CARE.

According to *Washington Post* columnist David Broder, "Nothing better displayed both the muscle and tactical planning of the opponents [of Clinton's health care plan] than the crushing of this forlorn bus caravan that summer. It was the crowning success of. . . . the conservative political interest group, Citizens for a Sound Economy."

Today, these tactics seem dated, overpriced, and relatively ineffective—in today's parlance, "Astroturf"—

relative to the real power of the decentralized community of freedom fighters that make up the Tea Party movement. Some things do remain constant, however, like the threats of massive energy taxes on carbon-based fuels and big, expensive, overbearing, government-run health care. Our mission remains the same, too: to defend the individual against the unjust encroachments of big government by empowering Americans to get involved and make a difference when and where decisions are made.

Years before the emergence of the modern Tea Party movement, FreedomWorks and a few like-minded organizations understood that taxpayer activism was alive and well. We were convinced that good ideas alone were not enough to win and that real social change came from the ground up, not the top down. We were reading Saul Alinsky, Barack Obama's mentor, a decade before it became cool. We also read *Dedication and Leadership* by Douglas Hyde and *A Force More Powerful* by Peter Ackerman and Jack Duvall. We understood that it is more important to come out of a legislative or regulatory battle stronger than you went into it than it is to simply win any particular skirmish. In the 1990s we translated this into an internal motto for staff and activists: "Winning by Building; Building by Winning."

At the time, however, these tactics were mostly employed by left-leaning groups—teachers' unions protecting a broken school system, radicals smashing windows at a World Trade Organization protest in Seattle, or public employees pushing for more wages and benefits and fewer hours worked. It seemed then that direct action, as leftists refer to it, was only used in efforts to expand the size and scope of government. The folks at FreedomWorks knew better.

It was no coincidence that FreedomWorks was at the center of the activism that followed Rick Santelli's rant. Indeed, FreedomWorks had facilitated some of the most impactful events just prior to Santelli's call to action and was already standing at the forefront of that first wave of political participation from the previously silent majority.

The very day of Santelli's outburst, FreedomWorks set up a Web site at IAmWithRick.com to give new activists access to basic tools and information. The site was an overnight success, earning tens of thousands of visitors within days of its creation. At the same time, FreedomWorks was flooded with calls and questions from first-time demonstrators eager to learn more about how to make their voices heard. Within just a few weeks of the rant, FreedomWorks helped coordinate dozens of taxpayer tea parties involving thousands of activists in places like Washington, D.C., Sarasota, Tampa, Jacksonville, Fort Myers, Tallahassee, St. Louis, Atlanta, Philadelphia, Rochester, Charlotte, San Francisco, Salem, and Sacramento.

A shocked mainstream media hurried to catch up. "The tea party concept has gained significant traction since Mr. Santelli's rant," the *New York Times* breathlessly reported. "FreedomWorks, a nonprofit group that mounts grassroots campaigns, has made Mr. Santelli the emblem of its efforts to oppose the stimulus, publishing his face on its home page and asking: 'Are you with Rick? We are.'"

Meanwhile, the *Saint Louis Post-Dispatch* offered an account from the front lines:

> *Critics of President Barack Obama's stimulus plan gathered beneath the Arch Friday to cheer speeches over a bullhorn and toss tea into the Mississippi River. Pleased with the turnout in 35-degree bluster, leaders said they had stolen a page from liberal tradition by taking to the streets with homemade signs. "If I had known this many people would show up, I'd have charged admission," said Bill Hennessy of Ballwin, the lead organizer. "We'll do this every chance we get until Congress repeals the pork—or we retire them from public life." Hennessy*

estimated that more than a thousand people showed up. There was no official count, but the crowd spilled across roughly one-fourth of the grand staircase from the Arch to Leonor K. Sullivan Boulevard.

FreedomWorks was also active on the cutting edge of technology that allowed a disparate group to quickly connect and plan events. A key tool that aided protesters in the earliest, most disorganized stages was a Google map that tracked activist events and allowed anyone with Internet access to find a group. The map was soon filled with virtual thumbtacks, a digital monument to the growing power of the movement. Staffers posted and shared key Twitter handles, offered advice on Facebook fan pages, and created massive e-mail lists of citizens who wanted to be informed of upcoming activities. The revolution, as it turned out, was not only televised. It was blogged, tweeted, texted, friended, and Facebooked.

SAMUEL ADAMS, COMMUNITY ORGANIZER

The spark that ignited the modern Tea Party movement was not just a question of bad economics—it cut to the core of basic American values of individual choice and individual accountability. Millions of Americans were still angry over the new culture of bailouts that had taken Washington by storm since the popping of the housing bubble in 2008 and they were just itching for a fight. They thought that candidate Obama would prove different, having run on a mantra of fiscal responsibility. Regardless of their limited choices at the ballot box, the American people were hungry for accountability, for the American way of doing things.

The entire founding enterprise, including America's Declaration of Independence from the British Crown in 1776 happened only because of the tea party ethos, the tradition of rising up against tyranny and taking to the streets in protest. Indeed, the period of American history leading up to the signing of the Declaration is the definitive case study in effective grassroots organization and the power of a committed, organized minority to defeat powerful, entrenched interests.

For any activist who fought in the trenches against Obama's hostile takeover of the health care system, the process that produced the Declaration will sound all too familiar: debate inside the Continental Congress was often dominated by lies, vote buying, and the influence of deep-pocketed business interests enjoying the favored treatment of the executive branch (King George III, that is). Does any of this ring a bell?

How did the advocates of liberty prevail over the entrenched interests and apathetic citizens that might have stifled the efforts of Thomas Jefferson and Benjamin Franklin? The answer, of course, is grassroots activism of citizens outside of the formal political process. The Declaration was radical in principle and revolutionary in practice—sweeping political change driven by a grassroots cadre of committed individuals armed only with their passion and their principles. Politics as usual did not stop them, and neither did lack of popular support. The political momentum for liberty was in large part created by the efforts of citizen patriots from Massachusetts, later joined by men in the other colonies. These so-called Sons of Liberty, led by a struggling entrepreneur named Samuel Adams—yes, the guy on the beer label—used targeted grassroots activism to undercut American support for British rule and create the political conditions that made ratification of the Declaration of Independence and the American Revolution possible.

Speaking truth to power was important, Adams knew, but nothing beat the power of grassroots activism. In the early 1750s Adams began recruiting activists to the cause of liberty, targeting men in taverns and workers in the shipyards and on the streets of Boston. His tactics often involved antitax protests under the Liberty Tree, a large elm across from Boylston Market. Tax collectors were hung in effigy and Crown-appointed governors mocked, belittled, and verbally abused. The Sons of Liberty organized boycotts of British goods and monopolistic practices that were de facto taxes on the colonists. Adams packed town hall meetings at Faneuil Hall, filling the room with patriots so that Tory voices were overwhelmed. Every oppressive new policy handed down by King George and the House of Commons was used to build the ranks of the Sons of Liberty. Taxes imposed by the Stamp Act of 1765, trade duties created by the Townshend Acts—each was an excuse to rally new recruits to the cause of American independence.

The most famous act of Whig defiance against the Crown—the Boston Tea Party—is now viewed as a tipping point in the battle for American independence. It had a profound impact on public opinion among the uncommitted population. It was not a spontaneous looting by angry tea drinkers but an operation carefully choreographed by Samuel Adams and the Sons of Liberty. When a Parliament-granted monopoly to the East India Trading Company dramatically drove up the price of tea in the colonies, Adams saw an opportunity to channel outrage into action. The "Mohawks" who emptied British tea into Boston Harbor on December 16, 1773, were his activists disguised by Indian war paint to protect their identities from Tory spies. Because property was not destroyed (other than the tea) and the ships' crews not harmed, the Boston Tea Party gave the Sons of Liberty broader public acceptance in the colonies.

The British response to the antics of the Sons of Liberty inevitably helped galvanize public opposition to British control. A series of political blunders coupled with the Port Act and the British blockade of Boston Harbor in 1774 ultimately led to the gathering of the first Continental Congress at Carpenters' Hall in Philadelphia.

The grassroots pressure organized by Adams was not reserved for King George and British bureaucrats exclusively. Adams also targeted Tory loyalists and "half-patriots" in the Congress. On one occasion, Pennsylvania delegate Joseph Galloway threatened to derail momentum for American independence with a last-minute proposal to settle disputes with Great Britain. Despite the well-articulated arguments of Patrick Henry and other patriots inside Congress, Galloway gained support among Tory and middle ground interests defending the status quo. Instead of arguing with his peers in the Congress, Delegate Samuel Adams organized outside, in the streets of Philadelphia.

In his classic account of the people and events that led to America's birth, A. J. Langguth wrote that the "crowds around Carpenters' Hall soon heard that a faction led by Joseph Galloway was bent on selling out their liberties. Galloway headed a powerful Quaker bloc, and yet he began to fear being attacked by a mob in his own home precincts." He backed down, knowing that Adams had organized his grassroots opposition. "He eats little," complained the Pennsylvania delegate of Adams, "drinks little, sleeps little, thinks much, and is most decisive and indefatigable in the pursuit of his objects."

Poor Samuel Adams. He was a pious man who resented the drinking and carousing of the privileged elite in favor with the British Crown in Boston. Today he is the mascot for a beer company. He was a tireless champion of individual liberty who dedicated his life to the battle for American independence. Today his tactics have been hijacked by leftist radicals hell-bent on tearing down the institutions that make our nation special. How did we lose our cultural heritage of grassroots activism to the big-government crowd? Isn't it about time we took it back?

LOOK OUT BELOW

To be sure, the brave men who authored and signed the Declaration were a unique blessing—smart, dedicated souls who define the word *patriot*. They put their principles first, and they put their lives on the line for the very idea of America. The final line of the Declaration pledging all to the cause, "our Lives, our Fortunes and our sacred Honor" was not empty rhetoric. The consequences of failure were dire and the signers knew that the ink on that parchment represented high treason against the British government, an act punishable by death. As the Declaration of Independence was sent to the printers for publication and distribution, signer Benjamin Harrison reportedly told one of his colleagues: "I shall have great advantage over you, Mr. Gerry, when we are all hung for what we are now doing. From the size and weight of my body I shall die in a few minutes, but from the lightness of your body you will dance in the air an hour or two before you are dead." These patriots were willing to risk everything to overthrow tyranny and establish a free and independent nation based on the sovereign rights of the individual. Many suffered greatly for their legacy.

These were the legislative entrepreneurs, the insiders, elected officials who must drive the process from within the legislative body. Without them, nothing happens. But it is equally fair to say that this defining act of courage by the founders would not have been possible without the men in the streets, the activists under the Liberty Tree, or the "Mohawks" who courageously scored one of the first public relations home runs in American history.

So don't go chasing that political unicorn of perfect leadership: good government based on the best arguments, implemented by selfless servants who miraculously leave their interests at the door when they enter public service. The founders knew that this was not possible, and that individuals given the monopolistic power of the state would inevitably sacrifice good government for more government power. More power, that is, unless kept in check. The best way to limit the natural self-interest of public officials and the influence of deep-pocketed interest groups is to restrain government and limit the incentives to manipulate tax, spending, and regulatory decisions.

BY THE PEOPLE, FOR THE PEOPLE

We accept the Tea Party brand because Santelli, perhaps unintentionally, reintroduced freedom-loving Americans to their roots and a fundamental tenet of our nation's fabric. "Tea Party" was a perfect moniker for the citizen rebellion that was brewing, and Santelli showed up at just the right time with a call to action that was remarkably suited to the cultural sentiments of a budding grassroots community.

America was founded by visionaries with great, entrepreneurial ideas: the primacy of the individual over the collective, a republican government constitutionally constrained within specific, narrow limits, and freedom of enterprise. Every small-government fiscal conservative knows and loves these ideals. But America was also founded, literally, on the revolutionary principle of citizen participation, citizen activism, and the primacy of the governed over the government. That's the Tea Party ethos. You find this referenced throughout the writings of the founders, and they are some of the best quotes from our history. Thomas Jefferson, in his correspondence with James Madison, wrote that "the people are the only sure reliance for the preservation of our Liberty." None other than George Washington, in his first inaugural address in 1789, reflected an appreciation for the importance of citizen participation and vigilance in defense of liberty that was at the heart of the American experiment. "The preservation of the sacred fire of liberty, and the destiny of the Republican model of Government," he said, "are justly considered as deeply, perhaps as finally staked, on the experiment entrusted to the hands of the American people."

In other words, Washington believed our newly minted government was dependent on the people to function properly. When did we all forget this part of the deal? When did we ever decide that the American citizenry could leave public policy to public policy experts, or, heaven forbid, to those astride the levers of power inside the federal government?

The Founding Fathers understood this. After thirteen years of struggle and a successful war for independence, they set out to write the rules of a new nation, a nation conceived in liberty. The Constitution delineated severe limits on government power, created checks and balances within the system to help maintain those limits, and outlined the rights of individuals that could not be infringed upon by the government. Instead of the government granting rights to the people, the Founding Fathers reversed the equation and had the people granting specific powers to the government to function in a few areas while allowing the people to exercise their unalienable rights.

Winston Churchill would later say that the American Constitution was the most profound act of political genius in the history of the world. It's hard not to view our nation's founding as a political miracle, but the Sons of Liberty knew better. It was the passionate participation of Americans committed to liberty and willing to show up in its defense that made the theory of individual freedom a political reality. While the founders did everything they could in the Constitution to establish a government that would preserve the rights of individuals, they were painfully aware that those freedoms would be protected only through the constant vigilance of the Sons of Liberty in succeeding generations. In other words, they understood that policy decisions are not driven by the best arguments, the best book, or a perfectly argued forty-page white paper. Policy decisions are driven by the people who show up, and the impact those voices have on the incentives of elected officials at the margin. Or as Samuel Adams put it, "It does not require a majority to prevail, but rather an irate, tireless minority, keen to set brush fires in people's minds."

3

BAILOUT

The straw that broke the back for a lot of [the people who marched on 9/12] was the statement from Bush where he said he had "abandoned free market principles to save the free market system" and then the actions that followed. Most had great concerns about the growth of government but had been too busy earning a living to pay much attention. TARP woke us up.

—Janet Marley, Kilgore, Texas

To fit their inaccurate narrative of the Tea Party movement as sore-loser partisans opposed to President Obama's agenda, many in the media suggest the Democrats' stimulus bill was the spark that lit this grass fire of protest.

They're wrong. The government expansion during President George W. Bush's reign provided the fuel. And it was his Wall Street bailout that ignited the firestorm we see today.

As the Democratic Congress led by Speaker Nancy Pelosi and Senate majority leader Harry Reid pushed through the Republican administration's bill crafted by former Goldman Sachs chief executive and then treasury secretary Hank Paulson, our predicament became crystal clear: we the people had lost control of our government. It was now the political class versus the American taxpayers.

Some thought they were voting to change this situation with the election of Barack Obama. Many who did are now disappointed that not only has he failed to bring such change, but he has doubled down on the bad policies of the Bush administration in favor of the political class—the political equivalent of pouring gasoline on a brush fire.

Many of us knew instinctually the bailout was wrong. We understood that in order for capitalism to work we need to be able to not only keep the potential gains from the risks we take but also accept the losses that may come. With profit comes the potential of loss. Many of us had a neighbor or heard about someone who had been living too high on the hog for too long and were wondering why we were now supposed to pay for it. And something just didn't seem right about the elected branch of government ceding so much money and power to an unelected bureaucrat.

We got it—our instincts were right. Unfortunately, many of the 535 people we sent to Congress didn't seem to get it. And they certainly haven't accepted their responsibility for creating the problem. As the spark that started what is now a historic movement, it is worth further examining just how bad a policy decision the bailout was, the economics behind the crisis, and some of the players who made it clear that the entrenched political class was one of the primary problems.

BOOM AND BUST

You don't have to look far to find the primary causes of the great financial panic of 2008. The origins of the housing bubble, the glut of dodgy mortgage-backed securities, the insolvency of the biggest players in the investment banking industry, and the stock market collapse that led to the creation of the Troubled Asset Relief Program and the Wall Street bailout of October 3, 2008, were all caused by bad government policy. These mistakes, some dating back a generation, created an inflationary bubble, severe economic misallocations, and a generalized "cluster of errors"—something economics tells us only happens as a result of government-created distortions. It was a tragic tale of systemic government failure. It was a morality play about the undue political influence by certain financial institutions on Congress and a reciprocal undue influence by key committee chairmen on key financial institutions.

One key culprit in this government-instigated debacle was the expansion of money and credit by the Federal Reserve system. Political pressure on the federal government's central bank to keep the money flowing is standard operating procedure in the nation's capital. Political jawboning for easy money knows no partisan divide. After the September 11 terrorist attacks, the Fed pursued an aggressive policy of money and credit expansion, eventually lowering its target rate from above 6 percent to 1 percent.

"Too many dollars were churned out, year after year, for the economy to absorb; more credit was created than could be fruitfully utilized," said economist Judy Shelton. "Some of it went into subprime mortgages, yes, but the monetary excess that fueled the most threatening 'systemic risk' bubble went into highly speculative financial derivatives that rode atop packaged, mortgage-backed securities until they dropped from exhaustion."

These mistakes were not inevitable, of course. And they were not unforeseen by sound economic thinking. The financial meltdown displayed all the characteristics of a classic government boom-and-bust business cycle generated by easy money and credit first described by Ludwig von Mises in the first half of the twentieth century. Mises was one of the leading proponents of the Austrian school of economics and had focused a substantial amount of his studies on business cycle theory.

Mises himself knew something about the personal costs associated with sticking to principle. A respected classical liberal scholar teaching in Vienna, Austria, he fled the rising tide of Nazi influence only to be largely ignored by leftist academia upon his arrival in the United States in 1940.

Loose monetary policy, according to Mises, corrupted the standard of value, distorted relative prices, and encouraged systemic malinvestments. The mania of easy money and credit generated errors in economic calculations and investment decisions. "True," said Mises, "governments can reduce the rate of interest in the short run. They can issue additional paper money. They can open the way to credit expansion by the banks. They can thus create an artificial boom and the appearance of prosperity. But such a boom is bound to collapse soon or late and to bring about a depression."

The inevitable correction may be painful, but attempts by government to inject new money into the economy to repair the real economic pain caused by the boom-bust cycle leads to more sustained pain, inflation, and economic stagnation.

Andy Laperriere, a financial markets analyst, raised these very concerns in the *Wall Street Journal* on March 21, 2007, a full year and a half before the crisis reached a boiling point. "Federal Reserve officials and most economists believe the problems in the subprime mortgage market will remain relatively contained," he stated, "but there is compelling evidence that the failure of subprime loans may be the start of a painful unwinding of a housing bubble that was fueled by easy money and loose lending practices." He warned:

Asset bubbles are harmful for the same reason high inflation is: Both create misleading price signals that lead to a misallocation of economic resources and sow the seeds for an inevitable bust. The unwinding of today's housing bubble is not merely an academic question; it is likely to inflict real hardship on millions of Americans.

Ted Forstmann, a private equity pioneer, also foresaw the financial crisis in a July 5, 2008, *Wall Street*

Journal interview. He pointed first to an increase in the money supply that changed the incentives of financial institutions. After making the same loans they normally would based on reasonable risk assessments and expected rates of return, banks "have tons of money left. They have all this supply, and the, what I would call 'legitimate' demand—it's probably not a good word—but where risk and reward are still in balance, has been satisfied." So "they get to such things as subprime mortgages, okay?" Forstmann says he does not "know when money was ever this inexpensive in the history of this country. But not in modern times, that's for sure." In fact, at the peak of the bubble, real interest rates were actually negative.

THE GREEDY LEADING THE GREEDY

The second half of this tale is the bureaucratic symbiosis between politicians and key players in the home mortgage financing industry, particularly Countrywide Financial and the government-sponsored enterprises (GSEs) like Fannie Mae and Freddie Mac.

Politicians aggressively enforced and expanded federal housing policy goals, effectively encouraging financial institutions to give loans to the unqualified. A social agenda of home ownership at any cost helped sustain an unsustainable housing bubble fueled with easy money and credit. After all of the bipartisan moralizing about promoting home ownership and after trillions of dollars of economic losses and personal destruction, actual home ownership quickly returned to its premania level. Indeed, a recent release by the U.S. Census Bureau found that the homeownership rate in 2010 was 67.1 percent—precisely the same rate as 2000. Political agendas aside, markets will eventually clear, leaving a trail of folly behind.

FreedomWorks and many others had warned Congress for years about the danger its twin monsters Fannie Mae and Freddie Mac were creating. As far back as March 9, 2000, we joined a coalition working "to shine the light of public scrutiny on the nation's two largest GSEs [government-sponsored enterprises], Fannie Mae and Freddie Mac," and set out "to educate our grassroots activists to the fact that these government-sponsored enterprises have exposed taxpayers to trillions of dollars worth of risk." By September 10, 2003, the House Financial Services Committee was forced to call a hearing on the potential danger presented by the GSEs, but nothing came of it. The comments that day by Rep. Barney Frank (D-Mass.) sum up the position of too many: "The more people. . . . exaggerate a threat of safety and soundness, the more people conjure up the possibility of serious financial losses to the Treasury, which I do not see. I think we see entities that are fundamentally sound financially and withstand some of the disastrous scenarios." Barney Frank, of course, went on to become the chairman of the committee responsible for overseeing the GSEs.

But we stayed at it and in 2006 named Fannie and Freddie among the Top 10 Welfare Queens in the country, warning, "Remember the S & L banking scandal from the 1980s? If the real estate market tanks today, taxpayers could be on the hook for billions once again. It's time for Freddie and Fannie to grow up and cut the cord from Uncle Sam's pocketbook." By early 2008 Congress was talking about a housing bailout. We responded with a Web site called AngryRenter.com for activists to voice their opposition to letting the housing market sort itself out.

When the housing bailout bill morphed into the Wall Street bailout, the downside for taxpayers grew. Again, we said let the market work itself out. That's what markets do. Once the price of something gets to the point where someone wants to buy it, they do—and no sooner. So long as the government tried to artificially keep housing prices up, there would be more houses for sale—more supply—than demand and the market wouldn't clear. The answer wasn't more government manipulation of the housing market. It was less.

The political process should have produced a rational, bold response, starting by unwinding the many government mistakes that created the housing bubble, repealing the various laws and regulations specifically designed to put people into homes that they could not afford and creating a more reasonable approach to how investment firms were monetizing various risks. We could scrap the Community Reinvestment Act and break up Fannie and Freddie, putting the pieces back in the private sector.

If liquidity and the availability of capital were an immediate problem, the tax on capital gains could be repealed, along with other tax provisions that punish savings and capital accumulation. A flat tax does all of this in one fell swoop. And finally, the various distortions in corporate accounting hurriedly drafted during previous legislative panics could be repealed, starting with Sarbanes-Oxley.

Most important, policies should have been designed to let the market work, not to attempt to artificially prop it up. The policy mistakes of the past dozen years could not be corrected without some pain. The only question left to decide was who would feel it. Markets do not work without allowing mistakes to be corrected and losses incurred. Bad actors on Wall Street or Main Street should suffer the financial losses produced by their bad bets and carelessness. Such losses, just like profits in good times, are vital to the functioning of markets. As then BB&T CEO John Allison noted in a September 2008 letter to Congress, "Corrections are not all bad. The market correction process eliminates irrational competitors. There were a number of poorly managed institutions and poorly made financial decisions during the real estate boom."

Sadly, Washington did not focus on the fundamentals. With a lame duck in the White House and a bitter election to determine the next president and political control of Congress, lawmakers focused on the expedient rather than the efficient. A major roadblock to change is the necessity for sitting committee chairmen to acknowledge their own mistakes and repeal their favored programs. So Congress defended the failing government-sponsored entities Fannie Mae and Freddie Mac while crafting a taxpayer-funded bailout to avert any turmoil in an election year. Even more surprising, when the Senate passed sweeping financial legislation in May 2010, reforms of Fannie and Freddie were specifically rejected despite the utter failure of these bankrupt entities and their role in the financial meltdown.

Government intervention into private enterprise of this magnitude always creates what monetary economist Gerald P. O'Driscoll Jr. calls "crony capitalism." This philosophy, he says, "owes more to Benito Mussolini than to Adam Smith. . . . Congressional committees overseeing industries succumb to the allure of campaign contributions, the solicitations of industry lobbyists, and the siren song of experts whose livelihood is beholden to the industry. The interests of industry and government become intertwined and it is regulation that binds those interests together. Business succeeds by getting along with politicians and regulators. And vice-versa through the revolving door."

Rather than address the problem, Congress asked the American people to fund the largest bailout of Wall Street in American history. The proposed solution to a problem created by easy money and easy mortgages was $700 billion in borrowed (or created) money to allow the government to buy dodgy mortgage-backed securities from failing Wall Street firms, the correct price of which Wall Street's best minds could not calculate. The bailout socialized a big piece of our private financial system, granting the U.S. Treasury secretary full discretion to dictate winners and losers in this reshuffling of assets. Rather than addressing the underlying problems, the legislation simply allowed the Treasury to prop up failing investment houses that took on risks they should have avoided and investments they should not have made.

THE PANIC BUTTON

It is remarkable how quickly both the formal and the informal constraints on government action either unravel or are simply ignored during times of crisis. Or, as Federal Reserve chairman Ben Bernanke so indelicately said during the height of the panic, "there are no atheists in foxholes and no ideologues in financial crises." Principle was replaced with the need to do something—anything—to respond to the economic emergency.

Who would have thought that a Republican president (claiming that he had "abandoned free market principles to save the free market system") would push for an unprecedented $700 billion bailout of Wall Street and later, Detroit? Who could have guessed that this wholesale abandonment of constitutional restrictions on the power of unelected bureaucrats would receive such a tepid response from a Beltway policy community that ostensibly existed to boldly challenge such bad ideas? Who might have imagined that a Democratic president, having voted for the same said bailout as a

U.S. senator, would seek to institutionalize the practice of private profit-seeking and socialized (i.e., taxpayer-funded) risk, in the process guaranteeing future bail-outs and bolstering the political culture of "too big to fail"?

This story has all the makings of a classic Greek tragedy, doomed to be repeated again and again. The lessons of the great financial meltdown of 2008 continue to be ignored by federal policymakers. Unabated, the unrelenting pursuit of these policies will create the conditions for another, bigger inflationary bubble that will continue to disrupt our economy, degrade the purchasing power of our currency, and subversively erode the wealth of working Americans.

On September 20, 2008, Treasury Secretary Hank Paulson released his department's "Legislative Proposal for Treasury Authority to Purchase Mortgage-Related Assets." Official Washington had joined the fray in full bipartisan panic over the collapse of residential housing prices, and with it, the potential meltdown of certain massive investment banks overleveraged with toxic mortgage-backed securities. Because of these events, the economy faced real and painful readjustments. But the real meltdown was actually happening in the government itself, where just about everyone in and around the levers of power, many of them culpable in the creation of the financial crisis, threw out any notion of constitutional restraint of government power and proceeded to make an unelected bureaucrat the de facto czar of the entire economy.

The Democrats who controlled Congress quickly embraced Paulson's idea of giving an unelected official with close ties to Wall Street complete, unchecked power and $700 billion to take charge of the situation. The original draft, only slightly modified in the final legislation, said that "the Secretary is authorized to take such actions as the Secretary deems necessary to carry out the authorities in this Act, including, without limitation . . ."

Bailing out bad actors who took too many risks was exactly the wrong approach. The media's story line coalesced around a tale of Wall Street greed, unfettered profits, and a lack of regulation that ultimately brought the nation to its knees. Unfortunately, this version ignored the key players responsible for the financial meltdown, including politicians, the Fed, and the politically well-connected investment houses on Wall Street that would ultimately reap the benefits of a $700 billion taxpayer bailout. Perhaps the victors do write history, but ignoring the lessons of the last speculative bubble comes at a steep price. One outcome may be institutionalized bailouts going forward and a continuation of "too big to fail."

TOO LITTLE, TOO LATE

When John McCain "suspended" his presidential campaign to come back to Washington in the midst of the crisis, he could have single-handedly killed the bailout and offered an alternative that focused on the bad actors, including Fannie Mae and Freddie Mac, and securing an unstable banking system, unwinding the bad institutions, and ending the massive wealth transfers of taxpayer dollars under policies of "too big to fail."

"I do not believe that the plan on the table will pass as it currently stands, and we are running out of time," McCain told the press. "We must meet as Americans, not as Democrats or Republicans, and we must meet until this crisis is resolved."

Instead, both he and Democrat nominee Barack Obama essentially rubber-stamped the wishes of Congress and the White House.

It was an opportunity tailor-made for the Maverick to stand on good policy and political ground by taking on both Wall Street's bad actors and the political corruption of the housing market. It was, we believe, a unique opportunity for the sinking Republican ticket to revive its standing with the American people and distinguish itself from a discredited Republican establishment. But that didn't happen, and the McCain campaign never recovered. Republicans were tarred with TARP, even though the entire Democratic leadership had carried the legislation, on their terms, to President Bush's desk.

This unconstitutional abomination has since morphed into a many-tentacled monster, devouring two of the big three American automotive companies and blurring the once bright line between good and bad financial institutions. TARP socialized risk and enshrined "too big to fail" by propping up bad institutions, and even forcing sound banks into the same shamed status of a TARP recipient in order to protect the bad actors.

Every major policy proposal since, from the government "stimulus" spending bill (official estimate: $787 billion) to Obama's hostile takeover of our health care system (official estimate: $940 billion), has been built from a spending baseline of $700 billion. Nothing less than that is considered "serious" public policy. Any attempts to restrain the natural desire of legislators to spend, the last pretense of fiscal restraint, died on the day TARP was signed into law.

Personally, we find it hard not to blame Republicans for much of our current predicament. The Bush administration, aided and abetted by many Republicans in the House and Senate, virtually erased any practical or philosophical distinction between the two parties.

Excessive spending, a new Medicare prescription drug entitlement, and a culture of earmarks—and the personal corruptions that naturally flowed from these addictions—destroyed the Republican Party's standing with the American people. This left many voters disillusioned and unmotivated, helping pave the way for the dramatic Democratic gains in November 2008.

The political response to the housing crisis and pursuant financial meltdown had a fundamental and irrevocable impact on the American economy, with politics and Washington advancing as markets retreated. Indeed, the massive government bailout turned the workings of the market upside down, transferring risk from Wall Street to the taxpayer. While doling out more than $700 billion to banks and investment firms, Congress has done little to avoid future financial meltdowns. The new mantra, "too big to fail," has effectively ensured the banks and Wall Street that the government will be there for future bailouts. More than ever, taxpayers are at risk, faced with trillion-dollar deficits, trillions in liabilities, and implicit and explicit guarantees.

We argued then and still believe that TARP gives unprecedented and unconstitutional powers to the secretary of the treasury with little to no checks or oversight. This concern has been confirmed, again and again, by TARP's brief, erratic path since enactment.

"Rather than making the policy choices necessary to guide the Secretary's discretion," a legal analysis prepared for FreedomWorks notes, "Congress has given the Secretary far-reaching power to intervene in the nation's economy and effectively to nationalize American businesses—upon the thinnest reed of statutory constraints. And in doing so, Congress has effectively chosen not to make law, but rather to make the Treasury Secretary the lawmaker."

CONCEIVED IN LIBERTY, IGNORED IN WASHINGTON

Some argue that the Constitution is a "living document" and binding government action to its literal words is a quaint idea. Resolving the financial crisis supersedes any constitutional concerns, it was argued during the legislative debate over TARP. But the constitutional constraints placed on government power are particularly relevant during times of crisis. Once liberty is taken, it is seldom returned. Drawing on their firsthand experience with King George, the framers made the exercise of power difficult by design to protect liberty, to force deliberation, and to ensure accountability.

To protect individuals from an encroaching state, the framers drew upon the ideas of John Locke and Baron de Montesquieu—the chief intellectual fathers of the separation of powers. The danger of a powerful executive is what led James Madison to warn in Federalist Paper 47 that the "accumulation of all powers . . . in the same hands. . . . may justly be pronounced the very definition of tyranny."

It took then secretary Paulson just three months to remind us of the timelessness of this wisdom. With the $700 billion bailout, Congress gave him the authority to spend up to a quarter of the government's budget with virtually no oversight. And with blank check in hand, Secretary Paulson almost immediately began to spend our taxpayer dollars differently than he originally said he would. Indeed, the TARP Congressional Oversight panel has been severely critical about the lack of transparency and unwillingness of the Treasury to provide more information with respect to the allocation of funds, with Elizabeth Warren, chair of the panel commenting, "The American people have a right to know how their taxpayer dollars are being used, and so far, they have not gotten the transparency and accountability they deserve."

The treasury secretary first funneled money directly to small and large banks and other institutions such as insurers and consumer lenders. Then he shifted the bailout's entire approach from purchasing assets to purchasing equity ownership stakes in troubled institutions. And in the most dramatic shift yet, with unambiguous disregard for congressional intent, the White House used over $17 billion of the funds to prop up failed Detroit automakers—after Congress voted against doing so. Paulson, upon changing his mind again, was blunt about his new czarlike powers. "While the purpose of [the TARP legislation] is to stabilize our financial sector, the authority allows us to take this action."

This is precisely the sort of action the framers intended to prevent when they deliberately separated powers between the branches of government: keeping the executive from exercising authority that runs contrary to the will of Congress. They intentionally gave the elected branch of government—Congress— the power of the purse and the power to write the laws. The executive branch is only supposed to execute those laws. They did this because, being elected, the legislature could be held more accountable by the voters, helping ensure it remained a government of, by, and for the people. We can kick a congressperson out of office every two years. We can't do anything about an unelected bureaucrat. Having lived under the rule of a monarch, the founders were all too familiar with the tyranny of the unelected.

In fact, the idea that the legislature could not simply give another branch its core responsibility of legislating was so foundational it was known by its own name: the nondelegation doctrine. The Founding Fathers knew they had to make this division of labor explicit because, being accountable to the people, legislatures would tend to want to hand off as many difficult and potentially controversial decisions as possible. And unelected bureaucrats, being human, would be happy to take as much power as others were willing to give them.

CUTTING OFF THE INVISIBLE HAND

As for the treasury secretary's desired role to become economic czar and CEO of the American economy, Nobel laureate F. A. Hayek's famous essay "The Use of Knowledge in Society" offers the best rebuttal. "*If* we possess all the relevant information, *if* we can start out from a given system of preferences, and *if* we command complete knowledge of available means, the problem which remains is purely one of logic," Hayek argued. "This, however, is emphatically *not* the economic problem which society faces. . . . The reason for this is that the 'data' from which the economic calculus starts are never for the whole society 'given' to a single mind which could work out the implications and can never be so given."

Hayek used this argument to dismantle the idea that socialist systems could supplant price discovery through the market process with well-meaning, smart bureaucrats. Markets, through their decentralized nature and their ability to incorporate the individual knowledge of time and circumstance, provide a far superior approach for coordinating the individual plans that drive economic growth.

Not surprisingly, the Obama administration's enthusiasm for the discretionary power of the program, originally scheduled to sunset in December 2009, has not waned. Treasury Secretary Timothy Geithner readily sent the pro forma letter to Congress required to extend the life of the program for two more years. The Democrats now in power seem intent on using TARP as a slush fund for various pet projects.

COALITION OF THE UNWILLING

The day after Paulson released his sweeping plan, FreedomWorks quickly connected with other free market groups to assess their willingness to fight. It was a surprisingly small group. But a principled few stepped up, notably Andrew Moylan of the National

Taxpayers Union. The Club for Growth and the Competitive Enterprise Institute also weighed in, and Dan Mitchell at Cato and the folks at Reason.com offered much needed policy support. The groups joined forces with a few free market Capitol Hill staffers who also were feeling remarkably isolated in their efforts to stop the massive government bailout.

Many groups that would traditionally fight such a massive outlay of taxpayer funds to irresponsible businessmen were sitting this one out, on the sidelines. Others would actually support the Republican administration's efforts. The diffidence of many of our usual allies was striking.

Much of the recalcitrance from right-of-center organizations seemed to be driven less by principle or substantive policy analysis than by a reluctance to challenge a Republican president—even one who had been a great disappointment on the key issue of fiscal restraint. Were some more concerned with maintaining a relationship with the Republican White House than promoting the principles of limited government? We did not know.

Another problem is that it is much easier to pick sides—Democrat or Republican—and go along with whatever the leaders say. This is a simpler path than trying to think through each issue based on facts and principles. The more years one spends in Washington, the more tempting this path of least resistance looks. Plus, there are a lot more jobs for party loyalists than there are for those of us who are constant thorns in the side of the establishment because we have principles— as if that's a bad thing.

Some free market grassroots groups were conflicted, holding public debates within their own organizations in an attempt to maneuver around various relationships with the Bush administration, politicians in Congress, candidates, and financial interests. They used the same flawed logic many on Capitol Hill used to convince themselves to support this giant government power grab: We must do something to address this crisis. This legislation is something; therefore, we must do this.

Days prior to the vote, FreedomWorks delivered our letter of opposition to the Hill—a "key vote notice." This is a powerful tool in the arsenal of grassroots groups. Organizations like ours issue these on almost all major pieces of legislation. They outline our argument for or against a bill and let the members of Congress know we will track their votes and let all of our members know exactly how they voted. It's one of the ways we help our members hold their representatives accountable and, we think, make our democracy function better by increasing transparency.

We were alarmed by how few of our erstwhile allies issued similar statements. Shortly before a vote there is usually a flurry of such letters, and it is usually many of the same groups taking the same position. We saw far more letters on far less significant bills.

The usually reliable Heritage Foundation ended up endorsing the TARP legislation. On September 29, 2008, Heritage vice president Stuart Butler and former attorney general Ed Meese coauthored a brief entitled: "The Bailout Package: Vital and Acceptable." "The constitutional questionability of some provisions is worrying, as is the centralization of power," the authors argued. "Nonetheless, the situation is so grave that we must take unusual measures now and accept some negotiated arrangements that remain very troubling, provided they are limited in extent and time and are not accepted as a permanent part of our government."

Perhaps some found the words of the Heritage Foundation in favor of the bailout more persuasive than our words against it.

Other groups decided not to get involved. How they could claim to be staunch defenders of limited government and not get involved was beyond us. But what surprised us the most, along with the Heritage Foundation's endorsement, was the endorsement of the House bill by Americans for Prosperity (AFP), an organization whose Web site claims they "engage citizens in the name of limited government and free markets."

AFP released the following statement on the Economic Emergency Stabilization Act (EESA), the bailout bill that included TARP:

AFP RELUCTANTLY SUPPORTS EESA

The Emergency Economic Stabilization Act is far from a perfect bill, although some of the worst aspects of earlier drafts have been removed, including the so-called affordable housing trust fund, the say-for-pay rules, and proxy access provisions. Still, the EESA does nothing to address the causes of the current crisis and it carries a frightening risk of permanently increasing the role of government in the economy, which could lead to worse market distortions and future crises. We shudder at the thought of Congress and bureaucrats controlling our country's credit markets. Interest group politics could be injected into every aspect of economic life. . . . The current crisis, however, is grave. There were other options that deserved greater consideration, but the choice today is the EESA or inaction, and inaction is not a good option.

It was one of the strangest, most intellectually tortured endorsements we have ever seen released by a self-described free market organization. While Americans for Prosperity's endorsement of the bill helped provide the political cover Republicans would need to switch their no vote to a yea, the statement was at least prescient. The legislation has failed to address the fundamental problems responsible for the crisis. In fact, at best, the program is a poor attempt to hold harmless bad actors accountable and delay the necessary corrections the economy requires. And the frightening risk of expanding the government's control of the economy has proved to be very real. The bailout program continues, and the Treasury Department has virtually unlimited discretion in its allocation of these funds.

The long, lonely campaign, first against the GSEs, then the housing bailout, and ultimately the Wall Street bailout, revealed just how few friends the cause of liberty actually enjoys inside the Beltway when it really matters. At each step there were fewer and fewer with us in Washington. But a growing, soon-to-be Tea Party movement had been awakened that would result in more allies than we had ever dreamed possible.

In the days leading up to the final House vote, many friends and fellow travelers in the free market movement—businessmen, Republicans, and alleged limited-government conservatives—all went out of their way to scold us for our philosophical intransigence. It was as ironic as it was frustrating to hear "free marketeers" lecture us on the importance of the bailout. Many of these same people worked for organizations that for years made their living by holding politicians' feet to the fire to make them take the "tough" vote for freedom. Yet when Congress was poised to vote on perhaps the most significant expansion of government power we may see in our lifetimes, some chose the easier path, the path of appeasement.

What good are principles if you can ignore them or alter them or bend them just a little to accommodate a difficult decision? Sometimes the principle of the matter demands that you stand up when no one else seems to want to do it. In times of crisis, when the details are in dispute or even unknown, values, principles, and an unbending understanding of markets matter more, not less. Principles matter most under the most challenging of times, and principles matter most particularly when they are unpopular or inconvenient.

SETTLING ACCOUNTS

Unfortunately, we are only beginning to understand how much was lost due to the Wall Street bailout. This massive intervention into the economy has fueled a big-government mentality that has continued through to the new Obama administration, which has proposed trillion-dollar budget deficits as far as the eye can see, massive "stimulus" spending packages, and a major intrusion into the health care market. The government remains a major shareholder of General Motors, and more recently appointed two members to the board of AIG, a private company.

Whatever one might have thought of the proposed use of extraordinary and unchecked executive authority, the actual implementation of the Troubled Asset Relief Program by both the Bush and Obama administrations is now giving every advocate of limited government permanent indigestion. Americans for Prosperity, which initially supported the TARP, has since removed any evidence of that position from its Web site and now rails against government bailouts along with the majority of Americans.

TEA TIME

When Treasury Secretary Paulson first announced the White House bailout plan on September 22, 2008, Matt Kibbe was quoted in *Politico* predicting that the "grassroots reaction is visceral and going to be big." The prediction turned out to more right than we could have hoped or planned for at the time.

But when we first opposed the TARP legislation, we had the terrible feeling of lonely martyrs fighting for an antiquated ideology of dead men—an ideology that had been conveniently embraced during good times but was now to be replaced with a pragmatic modus operandi of doing something, no matter what that "something" happened to be. Being a lonely martyr is a bad strategy—the last possible strategy—and it comes with an awfully dark feeling to know you are going to lose a fight you can't possibly afford to lose. But we marched on with the company of the few allies willing to step in front of the Treasury's runaway freight train.

Despite our hopes that grassroots citizens could rise up and stop this policy malfeasance, the speed with which Treasury Secretary Hank Paulson, Representative Barney Frank, and Speaker of the House Nancy Pelosi hoped to ram the bailout through Congress left us pessimistic that opponents had time to mount an effective opposition.

But then our phones started lighting up with calls from outraged activists across the country. Delay was our immediate goal, as it would allow for more transparency and clarity about the proposed actions contained in the evolving—and murky—legislation. The Treasury Department and the Democratic leadership in Congress were moving quickly, before grassroots opposition could potentially kill the bailout.

At FreedomWorks, our online team worked overtime to build a new grassroots protest site at NoWall StreetBailout.com to channel taxpayer anger back to politicians inside the Beltway. In just a few days, the Web site's petition gathered more than sixty thousand activists—a ready and willing Internet army set to mobilize against the Paulson bailout. Our full network of hundreds of thousands of activists had kicked into gear as well. Hundreds of volunteer team leaders across the country set up meetings with their local congresspeople and senators and started phone trees with their local grassroots networks.

Inside the Beltway, we were working with our allies on the Hill to come up with constructive alternatives and amendments, but we needed more time. The political process is at best a blunt instrument for use in healing an ailing economy, and it is typically true that good public policy will be ignored until absolutely every other option is taken off the table. Our goal was to kill the bad idea first—an indiscriminate bailout that rewarded risky, sometimes dishonest, and immoral behavior—to force government to prudently address the real threats faced by capital markets, the banking industry, and the entire economy.

A few stalwart fiscal conservatives in Congress also bravely stepped up and stood against the plan to socialize major Wall Street investment banks, led by Rep. Mike Pence of Indiana. Pence had come out against the plan early, arguing that "economic freedom means the freedom to succeed and the freedom to fail. The decision to give the federal government the ability to nationalize almost every bad mortgage in America interrupts this basic truth of our free market economy." At the time he stood nearly alone among his colleagues, and he took tremendous heat for his principled stand, but he was quickly joined by other principled members including Jeb Hensarling, Marsha Blackburn, Jeff Flake, and Tom Price in the House; and Jim DeMint and Jim Bunning in the Senate.

The Friday before the first House vote, the *Wall Street Journal* reported that "lawmakers say they have received hundreds of calls and e-mails in recent days, almost uniformly against the idea of giving the government the power to buy billions of dollars in distressed assets to keep the financial system afloat." Before the first vote in the House on Monday, FreedomWorks staff had delivered tens of thousands of petitions to congressional offices. Members reported that phones were ringing off the hook, and opposition was virtually unanimous.

All of this was encouraging, but as involved as we were, we had no idea just how big, and how visceral, opposition to the bailout actually was in America. Grassroots activism was beginning to take on a life of its own. People were getting up off the couch, pushing away from the dinner table. It was time to stop yelling at the TV. It was time to do something.

On the Monday morning before the first House vote, Dick Armey delivered a long, thoughtful letter to each House member. It read, in part:

> *The difficult question each of you faces today is simply this: Do you believe that the political process, having produced many of the perverse incentives that resulted in our economy's current predicament, can solve these underlying distortions by essentially doing more of the same? I believe the answer to this question is unequivocally NO.*
>
> *As an elected official who took the oath of office swearing to defend and uphold the Constitution, should you today feel a greater allegiance to a president, or a political party? I believe that answer is, emphatically, NO.*
>
> *This is a big vote, one likely to be studied and second-guessed for decades to come. With an*

*understanding of the intense political pressures
each of you face in this tough election year, I ask
you to oppose this bailout. . . .*

*As a public choice professor, I used to begin
class each semester with Armey's Axiom number
one: "The market is rational and the government
is dumb." Those quick to call for more regulation
forget the power of markets and refuse to acknowledge government culpability in the current mess.
Time and again, governments the world over have
attempted to outsmart the market and the current
legislation is no exception. And time after time,
markets respond, toppling the best-laid government plans as they move to correctly price the
underlying assets in exchange.*

This letter was being passed around the House
floor several hours before the vote by members of the
Republican Study Committee, a conservative caucus
of about a hundred Republicans then chaired by Jeb
Hensarling.

THE FAT LADY SINGS

On September 29, we gathered at FreedomWorks'
headquarters to watch the final deliberations on the
House floor. The staff crowded into an office, watching

C-SPAN on a small television. With the sound off, we followed the ticker of yeas and nays on the screen. We were tired and resigned. The fight was over and we considered the vote a formality, a fait accompli.

But the vote tracker began to tell an important and surprisingly different story. We started to yell and clap and cheer for the nays as opposition to the bill grew. It kept growing. Turning up the sound and seeing the final tally, we were as surprised as anyone that the House bill failed.

Against all of our expectations, at 2:07 P.M., the first legislation was defeated 228–205, with 133 Republicans voting against their president. FreedomWorks had been doing everything we could to stir up opposition to the bailout bill, but our splash of cold water had been consumed by a grassroots tsunami that crested over the floor of the U.S. House of Representatives.

In retrospect, September 29 is clearly the day the Tea Party movement was reborn in America. You can almost hear Samuel Adams calling us into action: "If ye love wealth better than liberty, the tranquility of servitude, than the animated contest of freedom, go from us in peace. We ask not your counsels or arms. Crouch down and lick the hands which feed you. May your chains sit lightly upon you, and may posterity forget that you were our countrymen!"

There was a massive wave of spontaneous grassroots outrage that rose up against the government's proposed actions, temporarily taking back the people's house from the political elite. While FreedomWorks, our tiny coalition of like-minded organizations, and a handful of true blue legislators toiled away, surrounded on all sides by the Beltway establishment, the citizens of America—for a few days at least—took their country back. The *New York Times* reported: "Americans' anger is in full bloom, jumping off the screen in capital letters and exclamation points, in the e-mail in-boxes of elected representatives in the nation's capital." We were told by our allies who work in Congress that constituent communications were 100 to 1 against the Paulson Plan. It was a shining, if all too brief, moment where grassroots America and the cause of liberty beat the Beltway establishment. While the bill would ultimately pass, it had stirred the passion of the grassroots freedom movement.

A few days after that glorious House vote, in what we now know was a harbinger for things to come during the health care debate, the Senate quickly porked up the bill to buy the additional votes needed to pass it and send it back to the House to be rubber-stamped. These "emergency" provisions included Section 503 of the act that, according to the official Library of Congress summary, "Exempts from the excise tax on bows and arrows certain shafts consisting of all natural wood that, after assembly, measure 5/16 of an inch or less in diameter and that are not suitable for use with bows that would otherwise be subject to such tax (having a peak draw weight of 30 pounds or more)."

The Democrats, the liberal apparatchiks at the Center for American Progress, the SEIU, and the Obama administration's partisan advocates in the media all love to ask the same question of the Tea Partiers: "Where were you when the Bush administration was violating the principles of fiscal responsibility, accountability, and limited government?"

The answer for many is, "On September 29, 2008, I stopped yelling at the TV, got up off the couch, picked up a mouse and the phone, and decided it was time to take America back from Washington."

As the Washington political establishment was about to discover, these newly minted citizen activists were just getting started. As the economy faltered and the government grew, this nascent group of activists began to forge the modern-day Tea Party movement.

4

WHAT WE STAND FOR

A lthough Tea Party activists come from a variety of backgrounds, they are united in a core set of beliefs. That is the inherent strength of the movement. When you have principle to guide your activism, you do not need an organizational hierarchy.

You'll notice this is a short chapter, and that is intentional. It just doesn't take a lot of words to say that we just want to be free. Free to lead our lives as we please, so long as we do not infringe on the same freedom of others. We are endowed with certain unalienable rights and delegate only some of our power to the government to protect those rights. Defenders of limited government understand that the U.S. Constitution lists the specific powers it delegates. If it's not mentioned, we retain that power. This is why the original U.S.

Constitution was only four pages. In a telling contrast, the recently proposed European Union Constitution was 100 times longer at 400 pages. That's because it *does* take a lot of words for rulers to tell unfree people which rights they will be given and how they must lead their lives. That's why Obama's health care legislation was more than 2,000 pages long.

Members of the Tea Party movement are focused on defending individual freedoms and economic liberty because one does not exist without the other. The overwhelming majority of activists are just responsible citizens trying to defend something they cherish: constitutionally limited government. This is a movement stirred into action not out of partisan bitterness but as a reaction to what they view as a government that has grown too large, spends too much money, and is interfering with their freedoms. When you speak with activists, no matter where you find them, four recurring themes inevitably become clear.

1. THE CONSTITUTION IS THE BLUEPRINT FOR GOOD GOVERNMENT

First and foremost, the Tea Party movement is concerned with recovering constitutional principles in government. Our nation was conceived in liberty and dedicated to protecting the inalienable rights of life, liberty, and the pursuit of happiness of the individual, not of the collective or groups of special interests. The miracle of the Constitution is the simple genius of limited government and its singular devotion to protecting individual liberty.

Our Founding Fathers designed a constitutional system based on private property and the rule of law to protect the individual from an overbearing federal government. An American's freedom is based on individual rights endowed by our Creator, secured by the Constitution. Among these are economic liberties that allow us to provide for families and pursue our own happiness. For more than two hundred years, American citizens have used their personal and economic liberties to pursue their dreams and provide for their families. Along the way we built a prosperous nation. American wealth was not an accident but a direct result of our freedoms.

Advocates of big government do not understand this. They take our freedom and prosperity for granted. In recent years we have watched as private property has been taken from families by the government and given to developers through the abuse of eminent domain. Under the health care legislation passed in 2010, the government mandated that all individuals must buy government-approved health insurance, whether they want it or not. The government should be concerned with protecting my liberty, not my liver.

The founders designed a government that was to do only that which was both right and necessary; the rest was to be left up to the states and individuals. It is simply the best organizational chart for running a society ever created. However, this division of labor only works if people mind their own business. The problem is that politicians and bureaucrats often do not know their limitations and make it their business to mind yours.

The Tea Party movement is asking to simply be left alone. The federal government should only exercise those powers we the people have delegated to it through our Constitution.

2. IN A FREE SOCIETY, ACTIONS SHOULD HAVE CONSEQUENCES

The second major theme running through the Tea Party movement is the call for personal responsibility. The founding documents built institutions that allowed for individuals to chase their dreams and be responsible for their own successes and failures. Tea Partiers value equality of opportunity, not equality of outcomes. For us, it is all about the rights of the individual over the collective.

These free and voluntary transactions are at the heart of our society. But when we are protected from the carelessness of our own actions, we tend to act foolishly. That applies to both business and individuals.

The Austrian economist Joseph Schumpeter noted that failure is an essential part of a functioning market economy. He described it as "creative destruction," as resources are constantly rearranged to their highest value. Without failure, you cannot have innovation. Without innovation, our standard of living stagnates.

For years, we have watched as people borrowed on credit cards and bought homes valued beyond their means. At the same time, businesses also borrowed and lived beyond their ability. Banks took disproportionately large risks and the big three automakers agreed to union demands for unrealistic employee benefits that they could not afford.

When it came time to pay the piper in the recession, we watched bailout after bailout. The system broke down as individuals and businesses were shielded by government from the consequences of their actions.

Those who had restrained themselves, saved, and budgeted were told their tax dollars would be used for the bailouts. And not just current tax dollars, but hundreds of billions of dollars in debt was taken out as well. Debt that will have to be paid out of future earnings.

3. THE FEDERAL GOVERNMENT IS ADDICTED TO SPENDING

The third theme found in virtually every Tea Party gathering is the conviction that the government is spending too much while unfairly expecting our children and grandchildren to pick up the tab.

Tea Party activists understand that the opposite of the invisible hand of the market is the invisible foot of the government. Every dollar spent by the government is taken from the private sector. Nineteenth-century French philosopher Frédéric Bastiat called this the "seen and the unseen," as the government points to everything it does with tax dollars, but what is not discussed is what would have been created by the private sector with that same money.

The economist Milton Friedman warned us that the true rate of taxation is government spending. With trillion-dollar deficits projected for years to come, we fear the greatest threat to our freedom and way of life will come from our out-of-control deficit spending. Inevitably the government will be free to either deflate our currency or impose catastrophically high taxes.

Today's spending is tomorrow's taxes. Whenever the Tea Party is protesting spending it is looking at the long term and protesting future taxation. Higher taxes degrade our standard of living, leaving citizens with fewer choices and fewer dreams.

4. OUR BLOATED BUREAUCRACY IS TOO BIG TO SUCCEED

The fourth dominant theme common to Tea Party activists is an understanding that the government has grown too large and invasive. The government can't control its own border or run the post office, let alone manage a bank or auto company.

The relationship between the private sector and the government is similar to that of a horse and a jockey. The winning combination is a strong and fast horse with a nimble and light jockey. When the jockey grows too large and the horse is starved, eventually the horse will collapse under the weight of the jockey.

The bloated public sector robs the private sector of much-needed capital investment. Capital is like fertilizer: when it's spread on the private sector it grows the economy; when it's fed to the government it grows more government.

Advocates of bureaucratic centralization have also invented a new nomenclature to support their policies, such as earned versus unearned income. Only the government can force you to hand over nearly 15 percent of your salary in Social Security, and then complain that Americans do not save enough.

And government begets more government. Whereas individuals in the real world have to live with the consequences of their decisions (unless they get a bailout), government does not because it can always get more money from the taxpayer. The only check on its growth is the ire of the citizenry. Government is also staffed by people who do not worry; they have the ultimate in job security. When a government program fails, the advocates of big government inevitably claim it failed because it was underfunded, not because it was a bad program.

Big government is driven by two audacities: (1) the presumption that people are dumb and don't know what's good for them, (2) people are corrupt and dishonest; therefore it is incumbent upon the government to take money and spend it on citizens' behalf. On the other hand, the Tea Party has trust in the practical genius of the American people to be responsible for making decisions.

Possibly the best illustration of the separate philosophies came in an exchange between former Republican

Senator Phil Gramm of Texas and a woman represent-
ing the education establishment. She wanted more government money and control of education and Gramm
wanted more parental control. According to Washington
folklore, the exchange went something like this:

> GRAMM: There is no one in Washington who knows
> and loves my children as much as I do.
> LADY WHO REPRESENTS THE EDUCATION ESTABLISH-
> MENT: I take exception to that Senator Gramm.
> I believe I do.
> GRAMM: Oh yeah? What are their names?

Every society has to find a balance between liberty
and security. Europe has chosen to put more weight in
security and subjects the individual to the needs of the
collective. The Founders intended to create something
different—a system that favors liberty. We value liberty highly and this commitment to the individual is
what makes America unique. As Ben Franklin once
said, "They who can give up essential liberty to obtain
a little temporary safety deserve neither liberty nor
safety."

We now find this commitment to personal and economic liberty being challenged by the government's
size. President Obama and Congress are looking to

"Europeanize" the United States through a legislative stampede of government control: the nationalizing of our health care system, cap and trade energy taxes, and aggressive unionization. Tea Party activists know you cannot have European-size government without European-size taxes and a corresponding loss of liberty.

The financial collapse of Greece should be the canary in the coal mine. Greece is the first Western country in recent memory to experience a sovereign debt crisis. In 2009 the Greek budget deficit was 13 percent of GDP and investors did not believe the Greeks could credibly get the budget under control. The Congressional Budget Office reported that the United States' 2009 budget deficit was 10 percent of GDP.

Clearly the United States is on an unsustainable path. In less than five years the interest on the debt alone will approach $500 billion. How long will it take before the market loses its faith in our ability to pay?

THE SOLUTION IS IN OUR HANDS

Soon after the adoption of the Constitution, Benjamin Franklin was asked what kind of government they had formed. He replied "a republic if you can hold it." Franklin understood the threats of special interests and enemies of liberty from around the world and within the fledgling nation.

While activists like Mary Rakovich and the others profiled throughout the book tend to share the common views described in this chapter, there is still great diversity among those who claim the Tea Party mantle. Make no mistake: disagreements exist and not all activists are in lockstep on every issue. But these individuals share a bond of bravery. In the face of criticism and doubt they walked out of their living rooms and into the street to fight for something they believe in.

The best way to start something, these novice protesters discovered, is to start it. Just get out there and do it. It was messy at first and each made their fair share of mistakes, but they learned on the job and formed powerful coalitions of private citizens that are beholden to no corporation, no union, no patronizing politician. Self-sufficient and self-sustaining, these networks live off the land and can strike at a moment's notice. They are the future of American grassroots activism.

Following her galvanizing protest in Fort Myers, Mary Rakovich could have returned to anonymity content in knowing that she had done her part. After all, few protesters could claim the kind of success she had achieved on their first foray into activism: coverage on national television, the ire of her opponents, congratulatory calls from friends and fellow citizens. Instead, she went right back to work.

Barely two weeks after storming the Harborside Center, Mary traveled to Phoenix. "Imagine my surprise when I found out the next stop on the president's bailout and stimulus road show was Arizona," she said. "I just had to do it again."

By the time President Obama arrived in Phoenix, the nascent Tea Party movement was beginning to gain steam. "In two weeks I went from being virtually alone to standing with a crowd of more than five hundred people. This is when I realized momentum was building."

After returning to Florida, Mary began working to prepare for the national round of tea parties scheduled for February 27, 2009. "I took my small but growing e-mail list, and with the help of a few new friends, we had a Tea Party on Fort Myers beach. More than forty people showed up, and this time the paper and local TV were there to cover it."

On April 1, a Tea Party event was held in Cape Coral and the number of attendees swelled to three hundred. "We didn't have a stage, but we brought a stump and a megaphone," Mary recalled. "You could only keep speaking as long as you kept getting thumbs-up from the crowd."

At the April 15 tax day protest in Fort Myers, more than two thousand citizen activists gathered to make their voices heard. "Now the press was really taking us seriously," Mary said.

As the health care debate heated up in Washington during the summer, Mary and her local group redoubled their efforts. "We had ten separate events outside of our local congressmen and senators' offices. We called, we wrote letters, and we marched. We were making our voices heard and getting more and more people involved in the political process. It was empowering."

After a summer of activism, Mary and her husband, Ron, decided to join their fellow citizens from across the country and participate in a march planned for September 12 in Washington. She had come too far to turn back now.

5

THE STATUS QUO LASHES OUT AT THE TEA PARTY

Something was going on outside the Beltway, out in America. People were indeed starting to speak out against those in charge, those who were mismanaging the public purse and abusing the public trust. And the establishment—the politicians, opinion leaders, the vested interests—started to notice. And before long, they felt threatened.

By April 2009 one leading indicator of the Tea Party's growing effectiveness was its expanding chorus of critics. They lashed out at the citizens who did not fit comfortably into the categories that normally defined politics as usual. The growing power and positive impact of this citizen revolt could now be measured in inverse proportion to the outrageousness of the claims about it and the attacks leveled against it.

It started with two left-wing bloggers—suffering from a deep state of denial—in a particularly vitriolic post at Playboy.com. Their claim, as implausible as it seemed, was that the entire protest movement against an out-of-control government was completely phony, a contrived conspiracy. They called it "Astroturfing." These were not real people with real concerns, they claimed, but a carefully orchestrated public relations campaign designed by partisan mercenaries to stop Barack Obama from achieving his destiny.

> *As veteran Russia reporters, both of us spent years watching the Kremlin use fake grassroots movements to influence and control the political landscape. To us, the uncanny speed and direction the movement took and the players involved in promoting it had a strangely forced quality to it. If it seemed scripted, that's because it was. . . . All of these roads ultimately lead back to a more notorious right-wing advocacy group, FreedomWorks, a powerful PR organization headed by former Republican House Majority leader Dick Armey.*

This paranoid fantasy about what they termed the "FreedomWorks mega beast" should never have seen the light of day. Indeed, *Playboy* quickly removed the post from its site. But not before the "Astroturfing" narrative was picked up by the liberal economist and Nobel laureate Paul Krugman. On April 11, 2009, he referred to FreedomWorks as the "Armey of Darkness" behind the Tea Party movement on his blog. Krugman repeated the story, virtually unchanged, in the *New York Times* on April 12. "It turns out that the tea parties don't represent a spontaneous outpouring of public sentiment," he wrote. "They're Astroturf [fake grass roots] events, manufactured by the usual suspects. In particular, a key role is being played by FreedomWorks, an organization run by Richard Armey, the former House majority leader, and supported by the usual group of right-wing billionaires. And the parties are, of course, being promoted heavily by Fox News."

Now everyone was repeating it. After all, it was in the *New York Times*. Three days later, Speaker of the House Nancy Pelosi parroted the fiction: "This [Tea Party] initiative is funded by the high end—we call it Astroturf; it's not really a grassroots movement. It's Astroturf by some of the wealthiest people in America to keep the focus on tax cuts for the rich instead of for the great middle class."

During the 2008 debate over President Bush's first tranche of what would be many bad taxpayer investments, totaling hundreds of billions of dollars, designed to prop up bad mortgages, Matt Kibbe was arguing with a reporter over activist participation with a FreedomWorks Web site called AngryRenter.com. Tens of thousands of very real people had signed our Angry Renter petition against the bailout, many leaving comments that reflected an economic wisdom woefully lacking in Washington. It was fake grass roots, the reporter claimed. "Well," Matt responded, "what do you consider real grass roots?"

"I'm conducting this interview," the reporter answered. "I don't have to define my terms."

In truth, the definition of *grass roots* sometimes depends on what side of the debate you are on. It depends on what the meaning of the word *is* is.

New critics started speaking up, sporadically at first, as willful ignorance of this rising tide of citizen discontent gave way to the realities. The social status of each voice seemed to increase as well, from the lunatic fringe to respectable scions of the Democratic establishment, as the status quo felt more challenged, more threatened, more at risk of losing power. The critics came from the left and the right of the political spectrum. They came from fringy bloggers and left-wing talk show hosts. Then they came from the mainstream media, think tanks, and elected officials.

Eventually, the biggest of big dogs started to bark: House speakers, former presidents, even The Man himself. The Obama administration itself would ultimately play the Astroturf card to explain the wave of opposition to the President's health care plan that arose at the congressional town hall meetings in August 2009. At an August 17 press gaggle aboard Air Force One, White House press secretary Robert Gibbs was asked to identify which special interests were opposing their hostile takeover of the health care system. "Dick Armey's group is out there," he responded, "actively getting people to go to town halls and yell at members of Congress."

PRESCRIPTION FOR GOVERNMENT CONTROL

The president was demanding that legislators rush a health care bill through Congress before the August recess. Another hurried, opaque, secretive legislative fire drill—replete with backroom deals cut behind closed doors with powerful pharmaceutical and health insurance interests. It's not difficult to understand why folks might get frustrated, even angry, that another trillion-dollar proposal, particularly on a topic as personal as your family's health care, was to be hurried to the president's desk and signed into law without a full, transparent, and honest debate.

It sure seemed like official Washington was willfully ignoring the wishes of its constituents, didn't it?

It seemed that way, because it was that way. In an undisciplined moment of inconvenient candor after the election, the new White House chief of staff Rahm Emanuel told the *Wall Street Journal,* "You never want a serious crisis to go to waste." There was a short window for the new Obama administration, "an opportunity for us to do things that you could not do before."

So the American people flooded town halls in August, prepared to ask tough questions about the Democrats' health care proposal. They showed up in droves because Washington was trying "to do things that you could not do before," just like it did with the stimulus bill in February and with the Wall Street bailout the prior fall.

As the Tea Party grew, the Democrats and their allies let their deepest insecurities get the best of them. The public was starting to understand! Accusations of Astroturf quickly gave way to uglier smears against the grassroots citizens who opposed big-government policies.

On August 10, 2009, Nancy Pelosi and House majority leader Steny Hoyer published an opinion editorial in *USA Today* entitled " 'Un-American' Attacks

Can't Derail Health Care Debate." "It is now evident," they argued, "that an ugly campaign is under way not merely to misrepresent the health insurance reform legislation, but to disrupt public meetings and prevent members of Congress and constituents from conducting a civil dialogue."

Before the August recess, Pelosi had let another of her own insecurities slip out, responding to reporters asking her about the failure to complete a bill by the end of July. "I'm not afraid of August," she said. "It's a month." Was she goading citizens into action—to rise up and prove her wrong?

Well, that's exactly what they did.

Once upon a time, the American Left celebrated grassroots participation in town hall meetings and a full-throated legislative debate as one of the very best American traditions—the vanguard of participatory democracy. This was the stuff of Norman Rockwell paintings. This was Freedom of Speech, circa 2009. They ought to have lauded these brave citizens for their willingness to get informed and challenge the political wisdom of the ruling class. That's part of what it means to be an American, right?

We've been involved in public policy debates long enough to know that when someone is losing an argument based on the facts, they try to change the subject.

Thus liberals' hysterical reaction to the rising public opposition to their hostile government reboot of the American health care system. Thus Democrats' hostile attacks on the citizens who overwhelmed congressional town hall meetings over the August recess. It was "un-American."

STEEPED IN BITTERNESS

Willful denial eventually gave way to ridicule and mockery, starting with juvenile locker room humor. On an April 14 report, CNN's Anderson Cooper opted for dirty jokes rather than his heavily branded expertise in investigative journalism, joking on-air that "it's hard to talk when you're teabagging" to a stunned audience. MSNBC's Rachel Maddow actually managed to say the words *teabag* and *teabagger* an amazing sixty-three times during a seven-minute segment on the April 15 tax day protests. This tasteless double entendre soon became good fun for the Far Left, seemingly a useful substitute for substantive policy arguments. Eventually, it too worked its way from the unserious fringes to the Oval Office.

On November 30, President Obama himself would pick up the derogatory phrase, telling Jonathan Alter that Republican opposition to his politically disastrous stimulus bill "helped create the teabaggers and empowered that whole wing of the Republican Party where it now controls the agenda for the Republicans."

Then the ridicule and mockery gave way to straight-up hate.

HITLER'S MUSTACHE

"They're carrying swastikas and symbols like that to a town meeting on health care," Nancy Pelosi said of health care town hall participants, implying that at least some of these citizen activists were Nazis. At the same time, liberal politicians and their allies were decrying the Obama is Hitler motif that they claimed dominated Tea Partier signage. The media loved to show photographs depicting protesters with a poster of Obama sporting Adolf Hitler's infamous mustache. The image made it into virtually every news story about the Tea Party movement. The sign did indeed appear at some protest rallies. But they never pointed out that the signs were carried by the leftist supporters of Lyndon LaRouche. These folks had nothing to do with the Tea Party movement, either organizationally or philosophically. It's hard to know what they want exactly (they used to protest with signs of George W. Bush with the same addition of facial hair), but it was easy to see they weren't Tea Party activists. At least to those who cared to look.

MyBarackObama.com, the Web site for the partisan advocacy organization that replaced Organizing for Obama after the 2008 presidential election ended, took the rhetoric to a whole new level. In an effort to generate counterpressure for Obamacare, on September 11, they issued a call to action for "Patriots Day, designated in memory of the nearly three thousand who died in the 9/11 attacks. All fifty states are coordinating in this— as we fight back against our own right-wing domestic terrorists who are subverting the American democratic process."

This post was quickly removed after Tea Partiers called them out, but the Democratic establishment continued to pursue this offensive narrative. Talk about out of touch with reality. The Democrats, suffering from a wicked case of rhetorical whiplash, went straight from calling Tea Partiers phony to senior White House adviser David Axelrod's inferring that they were somehow dangerous. "I think any time you have severe economic conditions," Axelrod told *Face the Nation*, "there is always an element of disaffection that can mutate into something that's unhealthy." When asked again if the tea parties were unhealthy, Axelrod hedged, saying "this is a country where we value our liberties and our ability to express ourselves, and so far these are expressions."

While it was generous for one of the president's top advisers to allow for a modicum of liberty and "our ability to express ourselves," we worried that some on the Left didn't take the First Amendment to the Constitution nearly as seriously as we had hoped. Were there now two sets of rules regarding our right to peaceably assemble and our right to petition the federal government with grievances?

Former president Bill Clinton would pick up the Democrats' "domestic terrorist" narrative in 2010 in another attempt to score some points against the Tea Party movement and FreedomWorks. This time the attack strategically arrived the day after our big April 15 Tax Day Tea Party in front of the Washington Monument on the National Mall. In a speech commemorating the victims of the Oklahoma City bombing plotted by Timothy McVeigh, Clinton was none too subtle:

I loved seeing that picture of him in the Post *today—the outline—Armey with his cowboy hat on. I remember when he called Hillary a socialist. . . . But what we learned from Oklahoma City is not that we should gag each other or that we should reduce our passion for the positions we hold,*

*but that the words we use really do matter be-
cause there are—there's this vast echo chamber.
And they go across space and they fall on the serious and the delirious alike; they fall on the
connected and the unhinged alike.*

You get the point.

PLAYING THE RACE CARD

Perhaps the most difficult and insulting attack Tea
Partiers have had to endure is the charge of racism,
first raised by one singularly angry comedienne,
Janeane Garofalo. "They have no idea what the Boston
tea party was about," she said. "This is about hating
a black man in the White House. This is racism
straight up. That is nothing but a bunch of teabagging
rednecks."

Former president Jimmy Carter later took this
charge from the lunatic fringe to the mainstream. His
comments arrived several days after the massive September 12 Taxpayer March on Washington in an attempt to explain the unpopularity of the president's
health care proposal. "I think an overwhelming portion of the intensely demonstrated animosity toward
President Barack Obama is based on the fact that he is

a black man, that he's African American," Carter told *NBC Nightly News.*

Do Democrats really believe that any person who disagrees with President Obama's policies is inherently racist? Of course they don't, but it's a great way to change the subject, to not talk about the fundamental problems with a government mandate that forces every American to buy a health insurance plan, the benefits of which are defined by the federal government, regardless of his or her income, age, health, or desire to do so.

President Carter obviously neglected to listen to any of the actual speakers at the event he targeted with his sweeping animus. There were Tea Partiers of every color on the stage on 9/12, including Deneen Borelli. As an African American, she is on the receiving end of more than her share of left-wing hate messages and racial slurs. Deneen has certainly paid her dues for daring to be part of this social network fighting for fiscal responsibility and limited government. Her rejoinder to Garofalo—"Hey, Janeane, my neck is not red"—got roars of approval from the sea of activists at the steps of the Capitol that afternoon. Indeed, her speech was so popular that she is now a regular fixture on the Tea Party circuit and a contributor on Fox News.

"The public is outraged about the president's policies—the spending, the budget, the deficit—not his skin color," Deneen said of Carter's claims. "It's easier for the Left to play the race card than address the public's legitimate concerns, but what the Left and the media are doing is damaging and dangerous. It's damaging because when everything is racist, then nothing is."

The charge of racism that the Left so casually throws around is like a nuclear weapon. It destroys more than its target. It tears at our social fabric and undermines Dr. Martin Luther King's mandate of a color-blind society. These phony charges do real damage to the cause of a civil, tolerant, and compassionate society. We are a grassroots movement made up of people who believe in individual freedom and individual responsibility. Racism and hate are inherently collectivist ideas. As individuals who believe in individual responsibility, we judge people as individuals, based on the content of their character, not the color of their skin.

From day one, the good men and women who have risen up in peaceful dissent against a government that is bankrupting America have been subjected to the worst kinds of ridicule, name-calling, and downright hate. Most of these attacks were partisan tactics motivated by the political ends of the attackers.

MAN BITES DOG

While Tea Party activists were absorbing abuse from the Left, there was "friendly" fire coming from the Right. Clearly, the Left has done its best to marginalize and dismiss the entire Tea Party phenomenon. While wrong, it is not surprising. What is more baffling, however, is that some from the right of center display a similar hostility toward the movement. Some in the comfortably established Republican old guard have also attempted to trivialize and ignore the importance of the Tea Party movement. Former Bush administration officials and think tank elitists alike have talked down the relevance and intellectual voracity of this new generation of grassroots activists, doing their best to take the steam out of a phenomenon they do not understand.

New York Times columnist David Brooks, for example, finds it hard not to look down from his lofty post as the resident "conservative" at the opinion page with a certain disdain for these activist rubes with their signs and bullhorns and their pocket copies of the Constitution. "Personally, I'm not a fan of this movement," he wrote on January 5, 2010, predicting that the "Tea Party tendency" could be the ruin of the Republican movement.

Fundamentally, his issue is with the philosophy of personal liberty and limited government embraced by these citizens. In March 2007, for example, just as his brand of big-government conservatism was destroying what remained of Republicans' standing with the American people, he argued against a return to the principles of Reagan and Goldwater. "There is an argument floating around Republican circles that in order to win again, the GOP has to reconnect with the truths of its Goldwater-Reagan glory days. It has to once again be the minimal-government party, the maximal-freedom party, the party of rugged individualism and states' rights. This is folly." People want security more than they want individual freedom, he argued then. Purposefully or not, the Republicans followed his advice. Political disaster followed.

As the Tea Party movement has ascended in popularity and grown in influence, Brooks's attacks have grown more pointed. In a condescending missive entitled "Wal-Mart Hippies," he argued that Tea Partiers have more in common with the New Left of the 1960s than with his brand of conservatism. The differences are trivial: sixties radicals "went to Woodstock," and the Tea Party "is more likely to go to Wal-Mart."

"The Tea Partiers have adopted the tactics of the New Left. They go in for street theater, mass rallies, marches, and extreme statements that are designed to shock polite society out of its stupor. This mimicry is no accident. Dick Armey, one of the spokesmen for the Tea Party movement, recently praised the methods of Saul Alinsky, the leading tactician of the New Left."

Besides the fact that we have studied the street tactics of leftists like Saul Alinsky, the decentralized grassroots network now commonly referred to as the Tea Party has very little in common with the "New Left," as Brooks claims. We are deeply rooted in the American traditions of individual freedom and constitutionally limited government. If it looks antiestablishment, that is because the political establishment has become completely and arrogantly dismissive of these timeless principles. And if that's radicalism, sign us up. The club is already populated with names like Jefferson, Madison, Washington, Franklin, and (Samuel) Adams.

Brooks claims that these pro-freedom protesters believe in "mass innocence." "Both movements are built on the assumption that the people are pure and virtuous and that evil is introduced into society by corrupt elites and rotten authority structures." No, it is mass self-interest we see as the human condition. The founders understood this and structured the institutions of our government specifically to protect against the deadly collusion of individual interests and unlimited power. We know that public officials act in their own self-interest, just like everyone else, so our strategy aims to support good ideas with the right political incentives.

Brooks also mocks Tea Party conspiracy theories dealing with big banks and corporations, among others. True, we believe that too many business interests conspire to use the power of the state to "compete" for market share, but our source is Adam Smith. You might call Smith a sixties radical of sorts; he wrote in the intellectual foment of the 1760s, when there was revolution in the air. "People of the same trade seldom meet together," Smith wrote in *The Wealth of Nations*, "but the conversation ends in a conspiracy against the public, or in some contrivance to raise prices." What do you suppose Smith would think about a $700 billion government bailout of banks authored by a former investment bank chairman turned government potentate, or trillion-dollar legislation mandating that every citizen buy the health insurance industry's overpriced product? That's right: it's a conspiracy, and it needs to be stopped.

Brooks doesn't like the new generation of small-government activists because, he claims, "they don't believe in establishments or in authority structures.

They believe in the spontaneous uprising of participatory democracy. They believe in mass action and the politics of barricades, not in structure and organization."

Maybe it's the decentralized, leaderless nature of the Tea Party movement that makes Brooks so uncomfortable. That's because freedom itself seems to make him uncomfortable. "Normal, nonideological people are less concerned about the threat to their freedom from an overweening state," he wrote in 2007.

Someone, Brooks seems to believe, needs to tell these people what to do, what to think, how to act in polite society. Is it possible that Mr. Brooks is the one, not Tea Partiers, with far more in common with those on the Left who desire order dictated from top-down structures, or as he puts it, "just authorities"? Someone needs to be in charge; that's what he is really saying. On this point, he might find common cause with Abbie Hoffman. Or Barack Obama, Nancy Pelosi, and Harry Reid.

Brooks and others on the right, such as David Frum, former speechwriter for President George W. Bush, who criticize the Tea Party movement for being too focused on the perils of too much government, appear more comfortable with the "go along to get along" approach to politics, vying for a seat at the table to make incremental changes to bad policies. This was very much the attitude that dominated Republicans in the House prior to the Contract with America in 1994. Accustomed to the perennial role of silver medalist, Minority Leader Robert Michel settled for negotiating improvements on legislation introduced by Democrats with very little vision for fundamental change. In effect, Republican leadership opted for a comfortable role as a permanent minority rather than a vehicle for change, settling for scraps at the Democrats' feast.

Today, some of the carping from the Republican establishment and the self-anointed thinkers behind it can be chalked up to self-preservation. They are looking down from the castle walls at the unwashed barbarians pounding at the gate. Senator Robert Bennett of Utah, whose bid for a third term in the U.S. Senate was rejected by Tea Party activists in his state, best sums up this worldview: "I'm convinced that the movement working against me is a movement of slogans, not solutions," he complained to the *Washington Post.* "Now I'm not a true Republican because I don't go on Fox and CNN and scream."

What is sending cold chills down the collective spine of the Washington political establishment is the now undeniable fact that the principles of limited government and fiscal responsibility have unprecedented political standing with the American electorate. There will be political consequences and those politicians out of step are losing their jobs.

This anxiety is echoed in the writings from the old guard of the political establishment. Michael Gerson, for example, another former speechwriter and an architect of compassionate conservatism for President George W. Bush, stated that much of what has recently come out of Tea Party activism is "a proposal for time travel, not a policy agenda. The federal government could not shed its accumulated responsibilities without massive suffering and global instability—a decidedly radical, unconservative approach to governing." In other words, big government is here to stay.

Perhaps what challenges the movement's many critics is the fact that the Tea Party does not buy into the traditional Left vs. Right debate. It is better framed as "big vs. small." It is a fundamental debate about the size and scope of government. Triggered by bailouts of irresponsible behavior on Wall Street, the Tea Party movement is first and foremost about fiscal responsibility—something that the political establishment across the Left-Right spectrum has failed to deliver. Trillion-dollar deficits and stimulus packages that only stimulate more deficit spending do not pass the commonsense test of kitchen-table economics.

THE THINKING MAN'S MOVEMENT

Other critics from the established think tanks seem more slighted that they were not consulted first, even painting the Tea Party movement as lacking tradition or intellectual underpinnings. Steven Hayward of the American Enterprise Institute, for example, wrote that "the conservative movement has been thrown off balance, with the populists dominating and the intellectuals retreating and struggling to come up with new ideas. The leading conservative figures of our time are now drawn from mass media, from talk radio and cable news. We've traded in Buckley for Beck, Kristol for Coulter, and conservatism has been reduced to sound bites." He laments the fact that "today's Tea Party has abandoned the intellectual icons that nurtured and expanded the conservative movement through the 1960s and 1970s. This reinforces the notion of the Tea Party as an angry, uneducated mob."

In fairness, Hayward has since warmed up to the Tea Party movement, but the mythology that Tea Partiers lack an intellectual underpinning pervades the ivory towers of official Washington. Brink Lindsey, vice president for research at the Cato Institute, best represents this view. While "opposition to Barack Obama and the Democratic Congress has sparked a resurgence of libertarian rhetoric on the right—most prominently in the 'tea party' protests that have erupted over the past year," real libertarians should not expect this uprising to translate into a fundamental realignment, Lindsey argued. "Without a doubt, libertarians should be happy that the Democrats' power grabs have met with such vociferous opposition. Anything that can stop this dash toward dirigisme, or at least slow it down, is a good thing."

"That, however, is about all the contemporary right is good for," said Lindsay, because of, "first and foremost, a raving, anti-intellectual populism, as expressed by (among many, many others) Sarah Palin and Glenn Beck."

One of the virtues of the online world we live in today is that mere citizens are no longer dependent on old-school institutions, like the U.S. Congress, or television networks, or the editorial page of the *New York Times*, for their information and their personal sources for good ideas. Like the Tea Party movement itself, access to information is completely decentralized by the infinite sources online. Like the discovery process that determines prices in free, unfettered markets, these informal networks take advantage of what philosopher Michael Polanyi refers to as "personal knowledge." Bloggers and citizen activists on the Internet now gather these bits of knowledge and serve as the clearinghouse for the veracity of facts and the saliency of good ideas.

Do Tea Partiers read? You bet they do, and with a focus and discipline fitting a peoples' paradigm shift away from big-government conservatism. We remember the woman who was one of the first to arrive at the front of the rally at the September 12 March on Washington. She draped a large white banner, almost as big as she was, over the crowd control barricade. It proclaimed succinctly: READ THOMAS SOWELL.

They have also read the Constitution of the United States, something many members of Congress probably cannot claim to have done. FreedomWorks for many years has distributed free copies of *The Law* by French free market economist Frédéric Bastiat to activists wanting to learn more about the intellectual underpinning of a free society. We also distribute Ayn Rand's *Atlas Shrugged* and other works by free market economists.

Cato executive vice president David Boaz, one of Lindsey's colleagues, points out that sales of books like *Atlas Shrugged* have skyrocketed recently. "It seems that Greenspan, Bernanke, Fannie, Freddie, Barney Frank, Bush, Paulson, Geithner, and Obama all created the objective conditions for an *Atlas Shrugged* sales bump," he wrote. *The Economist,* also cited by Boaz, makes a similar argument about Rand's new readership. "Whenever governments intervene in the market, in short, readers rush to buy Rand's book. Why? The reason is explained by the name of a recently formed group on Facebook, the world's biggest social networking site: 'Read the news today? It's like *Atlas Shrugged* is happening in real life.'"

The same is true of F. A. Hayek's *Road to Serfdom,* the Nobel laureate's short book warning about the perils of government-planned societies. The definitive edition of this book reached the number one spot on Amazon.com's sales rankings after Glenn Beck discussed the economist on his popular Fox News show.

These, of course, are not new books. Reading them, on the other hand, is a whole new generation of eyeballs. What has also changed is the addition of a few new classics to the activists' education. For instance, Saul Alinsky's *Rules for Radicals*—the original protester's handbook from the 1960s—has found a surprising readership among tens of thousands of Tea Partiers. This transition may provide some insight into the political establishment's unease with the Tea Party movement. It is a movement devoted to change, not simple academic debate. The Tea Party movement understands the tenets of the philosophy of freedom and is working to change the political landscape so those principles can be put into practice.

Rather than cause for alarm, this newfound activism should be welcomed and embraced by the traditional Right. In many ways the important intellectual arguments have been won by small-government conservatives. There are few big questions that remain unanswered by scholars. The dangers of excessive taxation, the threat posed by an activist monetary authority, and the importance of institutions and incentives for politicians and bureaucrats have all been examined by generations of scholars, and there are many who continue to examine these issues today. This body of work has created the basis for a political framework for a new limited-government movement.

But in the great debates over the expansion of the state in the twentieth century, these ideas were often ignored as politicians expanded the reach of the state and imposed new burdens on the taxpayer. From Roosevelt's New Deal through the creation of the Great Society by Lyndon Johnson, critics were vocal and insightful in the assessment of the potential threats to liberty. Even the recent health care debate saw its share of thoughtful and rigorous analysis by free market critics. Yet in most cases, the state expands and liberty contracts.

THE POLITICAL ECONOMY

The Tea Party movement adds a welcome addition to the fundamental debate over the size and scope of government: grassroots activists armed with the intellectual arguments they need to make a difference in political debates, not just scholarly discussions. What is happening is a dramatic increase in the physical infrastructure and on-the-ground personal politicking that can turn ideas into action. The new generation of limited-government advocates has its share of scholars, and the Internet provides an even wider audience for good ideas. But unlike earlier generations, the new generation has the muscle to make things happen in the political arena.

While standing for the right ideas and values is vitally important, it is naive to think that politicians will do the right thing simply because a proposed policy will benefit the general citizenry, creating the conditions for economic opportunity and individual prosperity for all. That's simply not how things work. If there was doubt about the proposition before, today it is painfully obvious that politicians in power often act in their own self-interest at the expense of the "public interest."

The "currency" that drives the political marketplace is fundamentally different from the private economy.

In the private economy, it is enough to have a good idea, identify a new product, develop it, and sell it to an identified (or created) customer base. In the market, entrepreneurship and competition determine outcomes. Returns and values matter and are ultimately determined by individuals making choices.

In the political economy, good ideas, philosophical values, and economic efficiency have little to do with how public policy decisions are actually made. The biggest error made by advocates of government planning, from Marx to Keynes to Obama, is the assumption that bureaucrats and elected officials possess both the detailed knowledge and right motives to be able to solve the economic problems of a nation. While microeconomics correctly assumes that individuals act in their own self-interest, every macroeconomic proposal for government intervention implicitly assumes that public officials act in the public interest, somehow suppressing their individual interests to the greater interests of society.

In reality, public choices are driven by the interests of those making the choices—the politicians who draft, promote, and vote on legislation; and the special interests that work to influence the political decision-making process. Politics is driven by the need to solicit new voters to the polls. Power (to tax, spend, and regulate) is used to consolidate those votes, and to buy more votes at the margin. The policy agendas of both parties are driven by this pursuit of votes and power.

As a result, those who do take the time to show up and actively participate in the policymaking process have a great deal of influence on the decisions that are made. Typically, this means that individuals looking for special treatment from government or a spending provision earmarked in an appropriations bill can dominate the policy conversation. They show up with an intensity that drives the legislative debate, because they stand to profit from the effort. While it is good public policy to limit the growth of government, the general public has little incentive to do much about it. Few dedicate the time and energy required to actively participate in the political process at a level that could influence the decisions of their representatives in government. For the average Americans, the costs of participating (knowing what is going on, showing up at town hall meetings, organizing like-minded citizens, and even voting) is much higher than the potential benefits.

Economists call this phenomenon "concentrated benefits and dispersed costs." You probably know it as "business as usual."

Given the way the world actually works, the only way to take power and money out of Washington is to create a powerful constituency demanding less. Government always goes to those who show up. The ideas of liberty need the political power that can be produced by an organized group of Americans committed to our values, trained in effective mobilization skills, and organized to drive policy at the local, state, and federal levels.

The Left figured this out a long time ago. Theirs is not an idea-based movement; instead, they focus on organization and power. As Saul Alinsky teaches, "change comes from power, and power comes from organization. In order to act, people must get together."

Today, the Tea Party has the power to change America for the better.

Many Democrats now sense this, and worry that they and the spokes in their political machine have gone too far in their aggressively hostile treatment of these concerned citizens. Headed into the 2010 midterm elections, some Democratic strategists have finally realized that the insults and smears and attacks targeting active, voting grassroots citizens is a very bad political strategy: "After a year in which Democrats have vilified the 'tea party' as an artificial grassroots uprising, as racist, and as potentially violent, several top Democratic consultants are saying that their own party would be wise to tone down its reactions to the small-government movement," *National Journal* reported on April 24, 2010. "There was some misjudgment at the very beginning of the process," former Clinton pollster Stan Greenberg admitted, in a dramatic understatement of the obvious.

Hank Sheinkopf, a Democratic political consultant in New York City, echoes Greenberg's call for cooler responses to Tea Party demonstrations and demands. "If not in 2010, it could be 2012 when we pay the price for their anger. . . . The attacks on them as 'tea baggers' and racists will inflame them and make the Democrats appear more elitist than they are."

Even President Obama now seems to acknowledge this when he calls for "a basic level of civility in our public debate." In a May 1, 2010, commencement address he said, "These arguments we're having over government and health care and war and taxes are serious arguments. They should arouse people's passions, and it's important for everyone to join in the debate, with all the rigor that a free people require. But we cannot expect to solve our problems if all we do is tear each other down. You can disagree with a certain policy without demonizing the person who espouses it. You can question someone's views and their judgment without questioning their motives or their patriotism."

Perhaps they are all reacting to consistent trends in public opinion polling, most notably by Rasmussen, that showed that despite the onslaught of attacks and smears from the Left, far more Americans identified with the views of the "Tea Party" (48 percent) than did with the views of President Obama (44 percent). Obamacare, which the rank and file of the Democratic legislative caucus had been promised would grow in popularity once people understood it, continues to lose public support. A strong majority want to see it repealed wholesale.

The promise of the Tea Party movement is its combination of the power of grassroots organization with the good ideas of freedom. The millions of patriotic citizens who are rising up—against a federal government that is spending too much and taking over too many things better left to free enterprise—will indeed take over the Republican Party and then the Congress. And then we will take America back from The Man, the term the New Left used to refer to the political establishment.

And that is why the status quo is lashing out.

ACHIEVING CRITICAL MASS

We'll never make it from here," said Freedom-Works press secretary Adam Brandon. On the morning of September 12, 2009, a slow-moving mass of people filled Pennsylvania Avenue from end to end, stretching as far as the eye could see. Police and parks department officials hurried past, walkie-talkies buzzing, unloading barriers and directing traffic. People from every state in the union, many of them marching with family and friends, filed down the street cheering and waving signs.

Already running late, we had hoped to quickly make our way to the stage to greet protesters and fire up the crowd. But this was more than a crowd—it was a sea of people who had flooded Freedom Plaza and the surrounding downtown area, and was splashing through side streets to fill every available space between us and the suddenly distant microphones. If we made it there at all, we would be late. Momentarily speechless, we glanced at each other in awestruck wonder. It was real. It was happening. And we were in the middle of it.

MARCHING INTO HISTORY

September 12, 2009, was a watershed moment for the Tea Party movement. Until that day, the pulse of the movement had been largely sustained through blogs, e-mails, phone calls, and any other tool that could connect activists in disparate locations. Tea Partiers, after all, tended to hold down full-time jobs and care for families. They were active in their home cities and counties but not known for crossing state lines or gathering en masse. Very few observers outside of FreedomWorks would have believed that a gathering on this scale was possible. . . . and there were many times we doubted it ourselves. The seeds of the march began long before anyone secured a permit or sent out an invitation. It began by encouraging activists to believe that anything was possible.

One of the books that is required reading at FreedomWorks is *A Force More Powerful: A Century of Nonviolent Conflict,* by Peter Ackerman and Jack

DuVall, a study of the history and tactics of the most successful nonviolent social change movements of the twentieth century. By far the most compelling chapter, one that we often distribute to activists during grassroots training sessions, describes the efforts of the brave civil rights activists who started pushing back against Jim Crow laws in the late 1950s. Intellectually, most opinion leaders and politicians in the 1950s and 1960s understood that government-sanctioned segregation and racial discrimination violated the promise of the Declaration of Independence. But real change did not happen until Dr. Martin Luther King Jr. and others took to the streets and mobilized tens of thousands of activist citizens to do the same.

Years of struggle by a relatively small group of fearless activists culminated with the March on Washington in 1963. This peaceful mass action was an iconic event that galvanized public opinion to push past the status quo of government-sanctioned discrimination against black Americans. The efforts of those first few souls who courageously sat in peaceful protest created seismic change and the needed momentum to move toward Dr. King's dream of a color-blind society.

By early 2009 we began wondering if could we do something similar in defense of individual liberty. We were talking about gathering thousands of protesters, maybe tens of thousands, in D.C. in a mass show of grassroots force. But was it even possible? We had never tried to pull off such a monumental task. We had only discussed this as a goal. Was it time to act? Could we do it? Opinions at FreedomWorks were divided and a healthy debate ensued.

It was now or never. If we didn't find a way to bring the voices of citizens together to put a stop to the crazed expansion of government being planned by Obama, Pelosi, and Reid, it would be very difficult to turn things around down the road. Somewhere there is a tipping point where burdensome government, dependency, unfunded liabilities, and oppressive tax rates institutionalize economic decline. Nations stagnate and die this way. Just look at some of the former economic powerhouses of "old" Europe.

IF WE BUILD IT, WILL THEY COME?

Now committed to the idea, we applied for the necessary permits to secure the September 12 date. We've often been asked why we selected that date. There were several reasons. First, we knew that we needed at least five months to plan such a massive event and that it would take a lot of time, money, media exposure, and word of mouth to drive attendance.

Second, we expected important legislation dealing with healthcare to be on the Senate floor in September. One of our members suggested September 12 because of his participation with the Glenn Beck–inspired 9/12 project.

Our first meetings with the National Park Service and Capitol Police were a bit tense. We didn't know what to expect or how they would respond to the idea. But as we went through the process we found them to be very helpful and easy to work with. The officials guided us through the red tape, answered our questions, and kept us on sound legal footing; they deserve a lot of thanks for making the event what is was.

But permitting was the least of our problems. We didn't have the money to fund such a big project. We didn't know how to organize an event on this scale. The march would test our staff and volunteers to the limit and require a great deal of learning on the job—with very little margin for error. But even if we found the money and navigated all the red tape, we still needed to get people to show up.

We didn't know it then, but that turned out to be the easiest part, because every participant shared the workload. The thousands of leaders who had risen up from their couches and kitchen tables over the previous weeks and months were building local networks that would be the heart and soul of the march.

They were finding one another online, meeting at rallies, and organizing via Facebook and Twitter and through other peer-to-peer networking tools that allowed these once isolated citizen patriots to join forces and work together. This decentralization of power and knowledge—tens of thousands of points of light for individual liberty—is the foundation for an organizational and philosophical revolution that would change the national debate in a profound, positive way.

A COALITION OF THE WORKING

Brendan, at FreedomWorks, along with Jenny Beth Martin of Tea Party Patriots and Darla Dawald of ResistNet, served as national coordinators of the Taxpayer March on Washington. Andrew Moylan of the National Taxpayers Union, a key ally in the battle against the TARP bailout, was a de facto fourth coordinator and a standout guy who was most interested in just getting the job done. FreedomWorks proceeded to assemble a large coalition of partner groups, creating a broad net of anyone who agreed on the basic principles of the movement: (1) a belief in individual freedom, fiscal restraint, and respect for our Constitution's

limits on government power; (2) we wanted a working coalition, one that understood the need to organize and take to the streets in defense of liberty. These
groups took seriously the founders' mandate of eternal
vigilance. Newly formed local Tea Party groups, Glenn
Beck–inspired 9/12 groups, and other local activist taxpayer groups made up the core of our coalition. They
were joined by a number of national groups, including
the Institute for Liberty, Tea Party Express, Smart Girl
Politics, Let Freedom Ring, Campaign for Liberty,
Free Republic, Leadership Institute, Ayn Rand Center,
and dozens of others.

We decided early on to forgo any big-name speakers,
in part because we didn't have the money. But we also
wanted the event to reflect the ethos of the Tea Party,
the leaderless nature of this spontaneous order. So the
emphasis was on the local leaders, the people who traditionally would not take to the stage and speak at the
podium, but labor anonymously, in the background, to
fill the town hall meetings, to organize church picnics,
to make community events or local political fund-raisers
successful. They never get the credit. Well, this would
be their day, their movement, their chance to speak to
the Washington political establishment, the national
media, and the American people. This also made the day
special, introducing a new group of leaders to America.

It was a paradigm shift that would make the old guard uncomfortable, inducing some carping and nit-picking from think tank elitists who should have been praising this grassroots rebellion. "No one asked our opinion," the ivory tower crowd seemed to intone from on high.

This "big tent" inclusiveness had one important caveat. The working coalition had to be united, like the Tea Party activists themselves, around the principles of limited government. This was a grassroots movement built upon a stage of ideas. We were motivated by good policy, not partisan politics. At the time, the sting of betrayal over the TARP fiasco was still fresh in our minds and it seemed like a perfect measure of one's commitment to good ideas over political parties.

It is quite easy to be on the right side of the ball in opposition to President Obama's outrageous stimulus spending spree when you are part of the Loyal Opposition. It is quite another thing to stand up against a Republican president perceived to be *your guy* and oppose him on good policy grounds.

As House majority leader, Dick Armey had occasion to say no to two Republican presidents named Bush, most famously in opposition to Forty-One's "read my lips" tax hike. It's no fun being the skunk at the garden party, but anyone can say "Yes, Mr. President."

So we made support for TARP a disqualifier. If the Tea Party movement was to grow and sustain itself as a social movement against big government past the present policy threats driven by Barack Obama and Nancy Pelosi, it had to maintain a fidelity to the values of individual freedom. No elected official or organization that had voted for or publicly supported the Bush/Obama/Pelosi/McCain bailouts would be included at the march. We politely declined inquiries from a number of friends, including a number of otherwise good congressmen and groups like the Heritage Foundation and Americans for Prosperity. This wasn't the easiest part of the job, but it was necessary.

Dick made an official announcement about the march at the April 15 tax day rally in Atlanta, Georgia, in front of an amazing crowd of fifteen thousand people who gathered on the steps of the state capitol. On April 17, after the massive protests across the country on tax day, we sent out an e-mail to all the state and local coordinators of Tea Parties we knew. One of the first local leaders to respond was Kellen Giuda from New York City: "You can count on my cooperation!" More feedback came in over the next few days, from Lorraine in Reno, Nevada, to Nikki from Mobile, Alabama. We had already been helping the folks in Atlanta with their tax day protest, and so we knew that Jenny Beth

Martin, Amy Kremer, and Debbie Dooley all loved the idea of a march on Washington. Other key local leaders started to respond with excitement and a willingness to help. Some of these first responders included Diana Reimer in Philadelphia, Robin Stublen and Tom Gaitens in Florida, and Toby Marie Walker in Waco, Texas. The march had passed from an idea to a vision we shared with thousands across the country. There would be no going back.

ON THE MOVE

In preparation for an official march announcement, we built a new Web site at 912dc.org for citizen organizers to use as a resource for logistics and as a place to meet up with other activists in their area. It was nothing fancy, but it did deploy important peer-to-peer networking functionality that helped eliminate FreedomWorks as the middleman in connecting local organizers trying to meet like-minded folks in their local communities. In practice, peer-to-peer networks create a multiplier effect that allows for exponential growth of local grassroots communities.

We used our Web site to sign people up to participate in the march, but the most important aspect it served was as a portal for a coordination of disparate people and their local knowledge. A great example of this was the organizing of local buses. Organizers from around the country would put down deposits on buses and then recruit riders to join the caravan and share costs online. On September 7, for instance, Suzanne from the Woodlands, Texas, posted: "We have started registrations for a second bus, so we still have space available. The bus leaves Thursday 9/10 at 10:30 A.M. and returns on Sunday 9/13 at 11:00 P.M."

Buses were organized this way all across the country, from Burlington, Vermont; from Montana, Idaho, and South Dakota; from Palm Beach Gardens and Jacksonville, Florida; from Bloomington, Evansville, and Fort Wayne, Indiana; from Portland, Maine, Joplin, Missouri, and Travers City, Michigan; all over Alabama, North Carolina, South Carolina, and Georgia; Zanesville, Ohio, Oklahoma City, Ephrata, Pennsylvania. Seemingly every state had caravans forming—buses were coming from all over the country, self-organized on our site and countless others.

We even had a delegation coming in from Hawaii. They had to fly, of course. Judging from their enthusiasm, however, they would have swum across the Pacific Ocean to participate on September 12 if necessary.

As we monitored the online chatter, we saw something new and special emerging, taking root and growing into something unexpected. In a word, this was going to be big. Publicly, we downplayed the numbers; better to be a pleasant surprise than a bitter disappointment. Internally, however, we talked about the possibility of breaking 100,000 people. We expanded our sound system to accommodate the big crowd. We also added more Port-a-Johns. To accommodate the expanded demands of the growing crowd, we created a fund-raising message at 912dc.org asking participants to pitch in. A donation of $45 would help pay for a foot of security barricade needed to manage the crowd during the march and around the stage. A Port-O-Let could be sponsored for $185. The JumboTron was funded in $1,000 installments. People who were coming and others who could not but still wanted to support the march quickly pitched in, allowing us to expand the facilities for the expected crowd.

THE SINCEREST FORM OF FLATTERY

Lefty bloggers had good fun with our fund-raising strategy, particularly as it related to on-site bathroom facilities. But at this point they had nothing left to throw at us except nastiness, smears, and name-calling.

The Democrats sensed that the momentum was on the side of those who were rising up in defense of liberty. They were getting anxious, and their nervousness led to mistakes. On August 18 former Clinton cabinet secretary Robert Reich called for a countermarch the following day, September 13, telling advocates of government health care and the euphemistically named "public option" that they had "to be very loud and vocal" if they hope to save their beloved vision of socialized medicine.

"We won't get a public option, or anything close to it, unless people who feel strongly about it make a racket," Reich told *Politico*. The "first step is to be very loud and very vocal: Write, phone, e-mail your congressional delegation and the White House. Second step: Get others to do the same. Third step: Get voters in Montana, North Dakota, Nebraska, and other states where Blue Dog Dems and wavering Senate Dems live and have them make a hell of a fuss. Fourth step: March on Washington."

That sounded like a challenge. Listening to the attacks leveled by Democrats against the citizens who showed up at town halls over the summer, you would have thought that the Left eschewed citizens who "make a racket," "a hell of a fuss," and anyone who is "very loud and vocal." Were we not just lectured to about such things? Was there a double standard?

Organizing for America, Barack Obama's lauded grassroots machine, did in fact pick up the cudgel and organize a counterprotest on September 13. Sophia Elena, a video blogger whose compelling footage of the Taxpayer March on Washington is some of the best on YouTube, captured the counterprotest the next day. All of 175 advocates of government-run health care attended. It was little noted and quickly forgotten.

The real power revealed on September 12 was the actual people who organized the march. True to the independent libertarian spirit of these decentralized protesters, there were tens of thousands of organizers for the march, and thousands of assorted caravans of buses from every corner of the nation. No speaker or attendee was paid a fee, and no transportation or hotel was subsidized.

EXPLOSIVE IDEAS

On the afternoon of September 11, 2009, staff and volunteers were gathered at FreedomWorks' headquarters at 601 Pennsylvania Avenue, scrambling to finish up the many projects that needed to be done. Campaign manager Nan Swift was working with the volunteers who were printing maps and directions, rolling T-shirts, and making protest signs. Brendan and Jenny Beth were working out a final order of speakers. Melissa Ortiz, a FreedomWorks volunteer, was working out the logistics of directing people around Freedom Plaza, along the route on Pennsylvania Avenue, and inside the barriers around the stage. Everyone else was on the phone, calling local organizers to bolster turnout.

And then Alberta, our office administrator, received the call: "There's a f—ing bomb in your building, bitch."

According to ABC News: "On the eve of what organizers call a 'Big Ol' TEA Party,' the Washington, D.C., offices of FreedomWorks were evacuated by D.C. Metro police on Friday afternoon after the conservative organization reported to authorities at 3:42 P.M. ET that it had received a bomb threat. Tens of thousands of anti–big government activists are expected in Washington on Saturday as part of a march on Washington being organized by FreedomWorks, a conservative group headed by former House majority leader Dick Armey, R-Texas. . . . D.C. Metro police has confirmed to ABC News' Jason Ryan that the D.C. Metro police had, indeed, evacuated the organization's offices. . . . Fifty volunteers were forced to leave the office where they were making calls encouraging people to come to tomorrow's event."

At the time the call came in, Matt Kibbe was being interviewed by Luke Livingston, who was shooting footage for *Tea Party: The Documentary Film*. Luke had traveled to Washington on a Tea Party bus from Atlanta with some of the activists he was following and filming for the documentary. When they knocked on the door to tell Matt and Luke's crew that the police were evacuating the building, the crew kept the cameras rolling. It made for a little drama in the movie, and Luke actually put together a short vignette that we played the next day. It was like a scene taken from a Hollywood production: police cruisers, sirens, bomb-sniffing dogs—we had it all.

It all happened so fast it was surreal. Certainly, some of the staff and volunteers were shaken up by the threat. Most of us were far more annoyed than scared. The incident, like the march the next day, was underreported, if acknowledged at all, by many liberal reporters in the media. Where was the outrage? Where were the scolding lectures to the Democrats and their liberal attack machine about political civility?

The best part of the incident was the resolve demonstrated by the volunteers who were making the march a reality. "What are they going to do," a volunteer asked while waiting to return to the building, "kill me, I guess?" The activists just shrugged it off and headed back up to the office as soon as D.C. Metro police was certain it was safe.

We would witness and hear this sentiment repeated over and over again among the activists of the Tea Party movement: Call us names. We will not take the bait. Ignore us. We will not stop. Threaten us. We will not back down. We love our country, we love our liberties, and this fight is too important.

When we finally worked our way up to the appropriately named Freedom Plaza on the morning of September 12, it all seemed worth the risk. It all seemed worth the work and the hassles and the threats. A beautiful sea of humanity greeted us as we worked our way to the stage.

NOW OR NEVER

We never did get to the stage. The plaza was too crowded with people. Tom Gaitens, FreedomWorks' Florida director, was firing up the crowd. "What do we want?" Tom asked the crowd. "Freedom!" The whole scene looked and felt like a carnival or a concert. Everyone was laughing and joking and enjoying the fact that they were participating in something that mattered. That day we all became a cohesive community of concerned, and now mobilized, Americans: 9/12ers, conservatives, Tea Partiers, libertarians, grandmothers and granddaughters, fathers and sons, independents, Republicans, and Democrats. You could find one of anything and/or everybody in the crowd that day.

We had planned to arrive at the stage on Pennsylvania Avenue at around 10:00 A.M., rally the gathering crowd with some quick comments, organize the various delegations by state, and march together down to the Capitol. Many of the groups that traveled in from across the nation had brought their state's colors to fly above the crowd in case things got chaotic. That way, people could find their local groups and march in an orderly procession down Pennsylvania Avenue at the designated time.

Billie Tucker and her First Coast Tea Party group from Jacksonville, Florida, had towed a massive replica of the HMS *Dartmouth* to Washington with them. The original *Dartmouth* was the first of the three East India Trading Company ships to dock in Boston Harbor in 1773 before the "Mohawks" emptied their cargo into the ocean in protest. The Jacksonville crew planned on "sailing" the *Dartmouth* down Pennsylvania Avenue when the police gave the signal.

Ready to lead the march was a contingent of Revolutionary War reenactors who planned to call the procession forward promptly at 11:00 A.M.

On the morning of the march, we arrived at Freedom Plaza. That is, we were trying to get to Freedom Plaza. In truth, we couldn't get anywhere near the area because that whole quadrant of the city was closed off by the National Park Service officers and D.C. city police who were trying to manage the mass of humanity that had started flooding into the plaza at 7:00 A.M. Freedom-loving activists from all over America were shutting down a good part of the city simply because of their sheer numbers!

So we walked toward Pennsylvania Avenue after police roadblocks stopped our cars far away from the center of the gathering. We were already behind schedule and scrambling to catch up with the citizens who had already claimed the 9/12 Taxpayer March on Washington as theirs. As we walked, wading through the growing crowd, it was clear to us that it was now their day. It was their march. It was their moment to petition their government for a redress of their grievances.

Everyone was in great spirits. We wandered around looking at all of the posters and flags. It really reminded me of a tailgate party at a football game. Groups would gather on one side of the plaza and cheering and singing would ensue, then it would start up on the other side of the plaza. There was a float parked along Pennsylvania Avenue that looked like a sailing ship hosted by a group called The First Coast Tea Party. A speaker on deck would rouse the group gathered there into frenzy and then they would play some patriotic music and when that was done they would start all over again.

—RON KAEHR, ALBUQUERQUE, NEW MEXICO

As we gathered at Freedom Plaza that morning, the crowd quickly overwhelmed the space and spilled out into the streets as the sun rose above Washington. The crowd became so huge, so fast, that the National Park Service officer in charge came up to Brendan Steinhauser at the stage and said, "It's time to go. You've reached critical mass." So, a full two hours earlier than scheduled, Brendan yelled into the microphone, "They are telling me that we've reached critical mass! If you're ready, I'm ready to march!" The roar of the crowd was incredible, and the mass of a million people began to make its way down Pennsylvania Avenue toward the Capitol.

It was an emotional moment. We were overwhelmed about what we were seeing. It is still overwhelming to think about it. We had always said that government goes to those who show up. We always knew that good policy would be considered inside the halls of Congress only when America beat Washington. We hoped that the citizens of our great nation valued freedom, free enterprise, and limited government as much as we did—not just because freedom is the only efficient way to allocate scarce resources and create economic prosperity for all, but because it is right and good and just to leave people free. And we always believed that the American people were with us, ready and willing to step up and take to the streets in defense of their liberties, just as the founders envisioned. Now we were seeing it with our own eyes: this beautiful mob of happy people—in the process of shutting down Washington, D.C.—was a dream realized for a small grassroots organization fighting for lower taxes, less government, and more freedom.

FACES IN THE CROWD

One of the great things about true grassroots gatherings is how different the culture is from a typical political event. This is not the place for partisan agendas, or politicians for that matter. There are none of the mind-numbing policy lectures typical of free market events. There are, in fact, no canned speeches carefully crafted from the best opinion polling data and focus groups.

There is lots of music and singing and chanting and talking and mingling and cheering.

The sheer size of the crowd could have been a logistical disaster because we simply were not prepared to accommodate and manage nearly a million people. Neither, apparently, was the city, the D.C. Metro service, the Capitol Hill Police, or the National Park Service. But it did not turn into a logistical debacle. Everybody seemed to go with the flow.

I loved seeing people of all ages marching! Elderly folks, lots of veterans, families with children, college kids, bikers, you name it! It was definitely a cross section of America, and as I looked at the crowd I was struck with the fact that these people aren't your normal protesters. They are normal people—the kind that make this country work! Most of us left jobs, school, family, and our normal responsibilities to go try to be heard and make sure the world knows we are out there and won't be silenced.

—MARILYN, DAVID, AND ANDREW TAYLOR,
CODY, WYOMING

By the time we got to the stage, set up in front of the U.S. Capitol, we knew that September 12 was one of the most important days for economic freedom and individual liberty we would witness in our lives. This wonderful mob of humanity quickly overwhelmed all of our carefully planned logistics and the city's transportation infrastructure, appropriating Washington, D.C., at least for that afternoon, to the cause of freedom.

THE NUMBERS GAME

The Taxpayer March on Washington was the largest gathering of fiscal conservatives in history. Nothing else comes close by comparison. Indeed, this was indisputably one of the largest protests of any kind ever in D.C. Martin Luther King's monumental March on Washington, by comparison, was 250,000 strong.

The size of the crowd was so large that it became the source of heated—and partisan—debate in the media. In fact, the debate started before the first activist arrived in D.C. A private memo released by the House Democratic leadership predicted a big crowd the morning of the march. "As you may know," the memo read, "Freedom-Works held a Capitol Hill demonstration yesterday that turned into an impromptu rally for embattled Rep. Joe Wilson. Now, based off of news reports and comments from leaders in the Tea Party movement, it looks like Saturday's event is going to be a huge gathering, estimates ranging from hundreds of thousands to two million people." This was the Democratic spin machine setting what it thought was a ridiculously high bar, presumably to ridicule the actual attendance the next day.

Bad strategy, Madam Speaker.

I was stunned by the number of people who were already there and more were streaming in. My excitement welled as I looked out over all these people. The energy and patriotism was amazing. There were flags everywhere, U.S. flags, Gadsden flags and military flags in all sizes. And despite the size of the crowd these protesters were polite and respectful. No pushing or shoving. I was also surprised at the respect for personal space. This was truly a peaceful protest. . . . these were my people! I could never have imagined this many people would be here.

—BILL HEERING, TERRYVILLE, CONNECTICUT

The London *Daily Mail* reported "as many as one million people flooded into Washington for a massive rally organised by conservatives."

The size of the crowd—by far the biggest protest since the president took office in January—shocked the White House. Demonstrators massed outside Capitol Hill after marching down Pennsylvania Avenue waving placards and chanting "Enough, enough!" The focus of much of the anger was the president's so-called Obamacare plan to overhaul the U.S. health system. Demonstrators waved U.S. flags and held signs reading GO GREEN RECYCLE CONGRESS *and* I'M NOT YOUR ATM. *The protest on Saturday came as Mr. Obama took his campaign for health reforms on the road, making his argument to a rally of 15,000 supporters in Minneapolis.*

NBC News estimated "hundreds of thousands." The silliest estimate came from MSNBC's David Schuster, who condescendingly tweeted that "Freedomworks [*sic*] says their dc demonstration attracted 30,000 people. Park police official says that is being 'generous.'" We, of course, never said that.

Our sound system, fully capable of reaching well over one hundred thousand people, was completely insufficient. According to the *Washington Post,* "Authorities in the District do not give official crowd estimates, but Saturday's throng appeared to number in the many tens of thousands. A sea of people surrounded the Capitol reflecting pool, spilling across Third Street and along the Mall. The sound system did not reach far enough for people at the edges of the rally to hear the speakers onstage."

We were caught off guard by the whole numbers game. The Park Police, who privately told countless participants that the crowd was easily over a million strong, no longer released official estimates after the Million Man March sponsors sued over an estimate that fell far short of the official name of their event. We do know that the peaceful demonstrators jammed Pennsylvania Avenue, seven lanes across, down to the Capitol, 1.2 miles in length, for more than three hours. The arriving crowd swamped the West Front of the Capitol and flooded down the National Mall and across various side streets—reaching to Independence Avenue to the south of the Capitol, and to Constitution Avenue to the north.

A grassroots movement that stands athwart Independence and Constitution seems just about right when you stop and think about it.

Charlie Martin, science and technology editor for Pajamas Media, did a crowd analysis that used time-lapse video footage to estimate the number of people who marched. The video shows the entire seven lanes of Pennsylvania Avenue filled along the entire path of the march for three full hours. Based on a very conservative pace and distribution, he concluded: "probably well more than 850,000 in the crowd. Which is a lot of people."

This, of course, does not include those who skipped the march and went straight to the stage for the rally. Then consider reports that traffic into the city was gridlocked. "We overheard people talking on their cell phones with friends in the outlying areas and learned that several roads and bridges were jammed for hours because of the high number of people trying to get to the rally," said marcher Ron Kaehr of Albuquerque, New Mexico. "From our vantage point at this time, the crowd stretched as far south as you could see. It was unbelievable! That's when I realized how large the crowd had grown."

The *New York Times* reported that "the magnitude of the rally took the authorities by surprise, with throngs of people streaming from the White House to Capitol Hill for more than three hours."

Conservatively, you can say that at least one million people showed up for the Taxpayer March on Washington on September 12, 2009.

A DAY TO REMEMBER

That morning we walked 1.2 miles from Freedom Plaza to FreedomWorks stage on the West Front of the Capitol surrounded by indisputable evidence that Americans uniquely treasure their freedoms and will rise up to protect them no matter the cost or inconvenience.

> *I will be fifty-one years old in October 2009 and drove 1,550 miles roundtrip to the D.C. Tea Party. This was my second protest in fifty-one years with the first being the Huntsville, Alabama, Tea Party on April 15 of this year. I found that my reasons for going were the same of most of the people I spoke with. First, we were there for our children's future and secondly, to save our country.*
>
> —MIKE GRUBER, MADISON, ALABAMA

As we approached the Newseum at Sixth Street and Pennsylvania Avenue, with its massive, seventy-four-foot-high Royal Pink marble facade carved with the forty-five words of the First Amendment to the Constitution, people began to read it aloud, in unison: *Congress shall make no law respecting an establishment of religion, or prohibiting the free exercise thereof; or abridging the freedom of speech, or of the press; or the right of the people peaceably to assemble, and to petition the Government for a redress of grievances.*

Some cried, and some cheered. Everyone smiled.

That so many people chose to come directly to the seat of our federal government—putting their lives on hold and adding to the strain on their family budgets— is remarkable enough. What was truly extraordinary was their unscripted, uncoached, and unrehearsed unity of purpose. We heard it over and over again in our conversations with individuals who made up this sea of fellow protesters: "I have never done anything like this before. But I have to do something; our government is out of control!"

Many in the media and liberal legislators alike try to dismiss these folks as simpleminded protesters opposed to taxes. In reality, they demonstrated a sophisticated understanding of economics: The real rate of taxation, as Milton Friedman argued, is the rate of spending. Government spending above current revenues must be paid for with borrowed money to be paid for with higher taxes in the future or government expansion of the money supply, which can only debase the currency and distort relative prices through inflation, making everyone poorer. Some had never heard of Friedman or Ludwig von Mises, or F. A. Hayek. But if you hold a job, manage a family budget, or run a small business, you just know these things. It's common sense. Imagine having even a fraction of this simple economic wisdom seated in the halls of government.

The huge, polite, focused crowd was like a fresh wind; 9/12 has changed me. I can no longer hide in my lovely little life and simply hope for the best. I can no longer hope my representatives and senators are looking out for our best interests. Freedom has always come at a price and to think we don't need to be vigilant and speak with our votes is lazy and naive.

—JENNIFER GALENA, OHIO

WHY WE MUST TAKE OVER THE REPUBLICAN PARTY

*Both parties seem to be more for big govern-
ment. . . . The Republicans need to learn that the
people they are running [for office] do not represent the views of the people.*
— SILVAN JOHNSON, FULTON, NEW YORK

Ever since Rick Santelli inadvertently branded this
grassroots citizen rebellion—fueled by disappointment in both the Republican and Democrat parties—
the media has increasingly confused the common noun
usage of the words "tea party" with the proper noun
Tea Party. "Citizens gathered in downtown St. Louis at
noon today to hold a tea party" soon gave way to "the
Tea Party is proving to be a political thorn in the side
of Democrats attempting to sell their health care legislation to reluctant members of Congress."

In many ways, it was a compliment to the fine men and women who had toiled for so long only to be dismissed by many in the media as phony, or worse, as the useful idiots of some shadowy syndicate of well-heeled corporate interests. At some point, the "Astroturf" caricature was abandoned by serious publications. Real reporters left their desks, actually went to some town hall meetings, and discovered, contrary to the Democratic National Committee's talking points, that these Tea Party folks were real, flesh-and-blood Americans.

Say what you want about their concerns, the Tea Partiers were energized and motivated and they were getting organized. That made them politically relevant. Political outcomes are almost always defined by the relative energy of various voting constituencies. If government decisions are defined by who shows up to advocate the passage or defeat of legislation, it is equally true that the voters who are motivated to show up on Election Day define who sits in public office and which party controls Congress.

There was growing recognition that the millions of people who self-identified with the Tea Party movement represented a potentially significant political force that needed to be covered by the press. Millions more people were starting to sympathize with the cause of fiscal responsibility that the Tea Partiers had brought to the forefront of the public conversation.

At some point, in the media's eyes, Tea Partiers became the Tea Party.

It is understandable why they got it wrong. Everybody gets it wrong. The Tea Party movement is decentralized. It is leaderless. No particular nominee, no executive director, no national chairman is in charge of this party. How this all works is literally lost on the political cognoscenti of the Beltway establishment. They simply cannot imagine something happening without the direct involvement of a designated authority, a boss, or, put in the currently stylish Washington parlance, a czar.

These, by the way, tended to be the same bright bulbs who so passionately argued that our country would simply stop manufacturing American-made automobiles unless the federal government took over General Motors and put the very best bureaucrats in charge of the business. Can't possibly make cars without a car czar, right? Have they ever wondered how on earth that loaf of bread ended up on the kitchen counter? Did the bread czar allocate just the perfect number of loaves in your sector to perfectly anticipate and meet your expected demand for a sandwich?

How could something that wasn't centrally controlled unfold in such an organized way?

PRINCIPLES, NOT POLITICS

The media needed to find a category to categorize us with; a box to box us into. They needed a neat and tidy story line that would fit the traditional partisan narrative of Us vs. Them, Left vs. Right, or Republican vs. Democrat. So they settled on Tea Party.

Say what you want about political parties, they are always intellectually and morally inferior to principles and good ideas. The sole purpose of a political party is to get candidates elected. Too often the candidate of one party could have just as easily run on the opposing political party's ticket. Political parties are empty vessels, adrift on tides that can shift with the winds of political opinion.

Principles, on the other hand, are different. Good ideas stand up to scrutiny. The right principles and the best ideas pass the test of time. They do not change based on the latest public opinion polling; they cannot be twisted like those dials on the machines experts use to measure the emotional intensity of a person's response during a focus group.

The principles of individual freedom, fiscal responsibility, and constitutionally limited government are what define the Tea Party ethos. They bind us as a social movement. And that makes the Tea Party better than a political party—something that can sustain itself the day after the first Tuesday in November, when all of the yard signs come down and all of the campaign volunteers go back to the daily routines of their normal lives. The Tea Party is a far more potent force for social change in America because it will sustain itself beyond the next candidate's election.

Be that as it may, when this band of citizen activists started to flex its collective grassroots muscle, the media's narrative shifted from characterizing the movement as a series of isolated events to describing it as a formidable—and potentially formal—political organization. They were searching for a measure to judge us. How many dollars are you raising? Who will run on the Tea Party ticket? When is the Tea Party convention?

The Democrats certainly hoped that this highly motivated constituency would create a third party that could split the more independent-minded fiscal responsibility vote from the Republican base. This would be of immeasurable help to swing Democrats facing difficult reelections. Embattled Democratic incumbents— forced to defend their votes for bigger budget deficits, more government debt on future generations, and an astronomically expensive and highly unpopular health care takeover—might eke past a majority of voter opposition if those votes were split between the Republican nominee and a third-party candidate supported by Tea Partiers.

Democratic political operatives looking for a silver lining in all of the public discontent consoled their ranks with the argument that this grassroots revolt was fueled by anger targeting all incumbent politicians equally. A February 2010 CNN poll seemed to bolster this line of reasoning. That survey found voter discontent was far greater than even 1994, when nine million new voters swarmed the polls and gave Republicans control of Congress for the first time in forty years.

"This is not a good year to be an incumbent, regardless of which party you belong to," said CNN polling director Keating Holland. "Voters seem equally angry at both Republicans and Democrats this year."

This might have been wishful thinking by panicked Democrats, but there was a faction of the Tea Party population who wanted to throw all of the bums out of office, hose down the Capitol, and start over with a clean slate. The notion had a certain emotional appeal and made for many of the most clever signs at the protests, tea parties, and marches. WE ARE HERE TO GIVE CORRUPT PUBLIC SERVANTS END-OF-JOB COUNSELING, one sign read. POLITICIANS ARE LIKE DIAPERS, said a placard held by a woman dressed up like Betsy Ross,

THEY BOTH NEED CHANGING REGULARLY AND FOR THE SAME REASON.

It was tempting to dump the lot of them, but it didn't make good sense to try to change things that way. After all, there were legislators who had fought for good policy and against bad policy even when it was considered political suicide to do so. Representative Mike Pence, for one, had risen in lonely dissent against his party's president and Treasury Secretary Paulson on the question of the Wall Street bailout in the fall of 2008. Many establishment Republican types pronounced that Pence's political apostasy would cost him his career. But for one congressman from Indiana—and for a number of his colleagues in the House and Senate who joined him to vote nay—that vote was defining. That vote was about principle, not party. Mike Pence would later become the House Republican Conference chairman, proving our oft-said adage: Good policy is good politics.

WHERE THE RUBBER MEETS THE ROAD

While planning the 9/12 Taxpayer March on Washington, we fought to include a select few elected officials including Rep. Pence, Sen. Jim DeMint, Rep. Marsha Blackburn, and Rep. Tom Price to address the crowd.

Some felt that the movement needed to avoid any affiliation with elected politicians, but we argued passionately that this grassroots army for freedom eventually needed to connect with like-minded individuals who actually held office. They may be hard to find, but there are principled public servants who will fight the good fight. These were the men and women we would need on the inside, in the legislature, writing the laws that would govern the next generation.

We have always argued that we needed to connect the energy of the Tea Party movement with a tangible strategy that would translate protests into policy, action into change. It was the same with politics. How would the Tea Party principles manifest themselves into more votes for the right candidates and tangible political outcomes? We needed to turn votes into victories.

FreedomWorks was hardly the only voice on this topic. Mark Tapscott, editorial page editor for the *Washington Examiner*, argued in September 2009 that Tea Partiers should "be figuring out how to channel this tidal shift in American public opinion into concrete results in next year's congressional elections" and that the "movement must identify and encourage like-minded candidates in both major parties." Erick Erickson of the conservative blog Red State echoed the sentiment, advising readers to "sign up for your local political party, encourage and support like-minded candidates and throw the kleptocrats out of office."

One evening on MSNBC's *Hardball,* Matt was asked by host Chris Matthews, "Would you guys knock off an incumbent Republican by going third party, because you know how the vote splits?" The prospect sounded tempting—after all, we were defined by good ideas, not by party affiliation. And every Tea Partier harbored deep disillusionment. The Republican Party was on double probation with all of us for its central role in expanding government in recent years.

Republican In Name Only (RINO) had become the label that disenfranchised Republican voters—the ones who had expected from their elected officials more than just party affiliation—used to tag politicians who talked the talk but seldom missed a vote to expand government spending.

Matthews was asking a good question: If we knew we could take a seat in Congress by running our Tea Party candidate, would we do it? He knew the right button to push, and he was probably hoping we would take the bait.

The answer was no.

FIRST, SECOND, AND LAST PLACE

In the real world, third parties don't win very often. And if you aren't elected to hold public office, you will never get a chance to enact laws that could benefit the fiscal futures of our children and our grandchildren. Part of the reason for this is our restrictive campaign finance laws. It is very difficult to raise the money needed to run a competitive election in the artificially small amounts dictated by the Federal Election Commission. It's much easier to start a successful campaign with a network of donors and a political party apparatus.

In practice this means that the few third-party candidates who do rise to national prominence tend to be eccentric billionaires on very expensive ego trips. Remember Ross Perot? In today's dollars, he spent almost $100 million of his own wealth, but he ultimately received zero Electoral College votes in his quixotic 1992 bid for the presidency. He did take 19 percent of the popular vote, making him the most successful third-party candidate since 1912. That was the year popular former president Theodore Roosevelt took a losing 27 percent of the vote as a Progressive Party candidate.

Like it or not, the two major parties currently control almost every single elected office in the country, from president to Congress, state legislatures, city councils, and town assemblies. We can safely predict, without fear of embarrassment, that either Republicans or Democrats will control the U.S. House and Senate in 2010 and the White House in 2012. Even when third-party candidates win—like former socialist Bernie Sanders from Vermont and former Democrat Joe Lieberman from Connecticut—it is under special circumstances and they will have to choose to caucus with one of the major parties anyway.

It is within the parties where would-be third-party candidates are making the biggest difference. Dennis Kucinich is perhaps the best-known "progressive" in Congress, but he's not elected on the Progressive Party ticket. He's a Democrat. And Ron Paul is the most successful libertarian in Congress; he's doing important work to rein in the Federal Reserve, but he's only ever been elected to hold public office as a Republican. When he ran on the Libertarian Party ticket, he lost.

As the Libertarian and Green parties' lack of representation at the federal level demonstrates, it is very difficult and can take a long time to build up the party infrastructure necessary to take control of Congress. Most important, every hour spent doing so would be one less hour than we can afford to lose in the fight to take America back from the big spenders in Washington, D.C.

When Dick Armey left his position as a university professor and chair of the economics department in 1983, he decided to run for Congress. At the time he had no party affiliation and zero party loyalty and had not paid a moment's notice to the practice of politics in almost twenty years. He considered politics a curious form of juvenile delinquency. Politicians, in his limited experience, were shortsighted and self-serving, like undisciplined children.

But if he was going to run for Congress he had to choose a political party. Which one? He chose the Republican Party because he had seen, on several important occasions during his own life, Republicans rise to meet his expectations. Barry Goldwater demonstrated a refreshing fidelity to the principles of freedom and a palpable respect for our founders and the entrepreneurial genius of our Constitution. Discovering his principled campaign for president in 1964 was Dick's political birth; he was a "Goldwater Baby."

When Barry Goldwater lost to Lyndon Johnson, Dick decided that principled people who stood for sound economic principles, personal liberty, and limited government could not possibly get elected. So he put politics aside and went to work, which he did for the next two decades. However, when Ronald Reagan won the presidency in 1980, he recognized that Reagan combined the same principles that had animated Barry Goldwater's 1964 campaign with resounding political success. President Reagan proved that you could win as a Republican by boldly offering voters a positive vision for America based on the principles of lower taxes, less government, and more freedom.

Today we certainly harbor our bitter disappointments with the Republican blunders that led to a Democratic takeover of Congress in 2006 and the White House in 2008. But those shining moments when the Republican Party embraced a national vision for America based on the principles of individual liberty and government restraint remind us that there is hope. There was hope with Goldwater and real change with Reagan. Republicans stood for something worthy again in 1994 with the Contract with America.

PROGRESSIVELY PROGRESSIVE

Today's Democrats have effectively silenced the fiscal conservatives in their caucus with a Far Left agenda of tax-and-spend policies that are anathema to a commonsense middle built around government restraint.

Today's Democratic Party is not much more than a coalition of special interests that want something from government. They want a program, an earmark, a regulation, favored treatment, or, if possible, a handout.

We have little hope that the Democrats could serve our positive vision for America. We despair over the state of that party. Moreover, it's too late—the "progressive movement" has already spent the past century taking over the Democratic Party.

Understanding how the Far Left hijacked the Democratic Party is instructive. The progressive movement in the United States began during the late 1800s and argued that government should get involved in making life "fair" and "equal" by mandating higher wages and shorter working hours, offering welfare, and curing social ills through alcohol prohibition, among other things. Progressives called for top-down solutions—using the force of the state to require everyone in society to behave as they desired—but politically, they used a bottom-up approach in local elections.

By the 1910s, progressives had become a substantial force. Buoyed by their inflated view of their own importance, they made a critical mistake that the Tea Party movement must avoid. Rather than taking over one of the two major parties, they decided to form a new party to run their own candidates. In 1912, the Progressive Party was formed and created a platform that went by a name that may sound familiar to you: the Contract with the People. They ran popular former president Theodore Roosevelt as their candidate for president. Even with one of the faces that ended up on Mount Rushmore as their candidate and decades of local campaign experience, they were only able to win 8 electoral votes to Woodrow Wilson's 435. Over the course of a decade, they put one governor, one U.S. senator, and just thirteen U.S. House members in office. That's 50 short of a majority in the Senate and 205 short of one in the House—which is to say, a colossal waste of time.

The Progressive Party faded away, but its ideas did not. By the 1930s, legendary progressive leader Saul Alinsky began organizing and training activists to be more effective. His goal was to organize a mass army of people from the community to pressure politicians into supporting "progressive" policies. Rather than spend time creating a new political party, he was going to spend his time more effectively and take over the existing structure. Alinsky—one of the original "community organizers"—is widely known as one of President Obama's greatest intellectual inspirations and the source of many of his organizing tactics. Alinsky's guide to being an effective activist is called *Rules for Radicals*. We've all read it at FreedomWorks and suggest you do, too.

The political landscape was changed as progressives influenced Democrat Franklin Delano Roosevelt, helping him pass the New Deal and select Supreme Court justices who argued for a "living constitution" that evolved with the times. Progressives continued their powerful influence over Democrats by mobilizing and campaigning for Democrat Lyndon Johnson and his plan to redistribute wealth through his "Great Society." With the New Deal and the Great Society, progressives succeeded in dramatically expanding the government's reach.

This massive burden of government produced the economic stagnation of the 1970s, and it seemed as though the progressive movement's political power had waned with the election of Ronald Reagan in 1980 and '84, the election of Bill Clinton from the more moderate "New Democrat" wing of the party, and the Republican takeover of Congress in 1994. But they never quit; they were quietly, continuously building their ranks within the Democratic Party.

In 1985, in the middle of Reagan's presidency, Democratic activist Ellen Malcolm formed the political action committee EMILY's List, which describes itself as "a community of progressive Americans" committed to electing women who shared their philosophy of expansive government. Malcolm noticed there were only two women in the U.S. Senate at the time and both were Republicans. To do something about it, she gathered twenty-four women in her basement to come up with an action plan to elect more Democratic women who supported their issues.

This network of women created the political powerhouse EMILY's List PAC. The list immediately started to produce money, which was sent to candidates the women liked. One year later, EMILY's List candidate Barbara Mikulski of Maryland became the first woman Democrat elected to the U.S. Senate.

In 1991 self-described socialist Bernie Sanders of Vermont formed the Congressional Progressive Caucus along with five other House Democrats. They set out to push, among other things, higher taxes, more welfare, and "universal health care." In 1992 EMILY's List's membership grew by more than 600 percent and raised $10.2 million to help engineer what political scientists now call "The Year of the Woman." Four new Democratic women were elected to the Senate along with twenty women to the House of Representatives, including several who joined the Congressional Progressive Caucus. EMILY's List had succeeded in becoming the first PAC to raise significant funds through small donations to change the political landscape.

In the late 1990s EMILY's List was joined by other progressive grassroots groups, including MoveOn.org, an early pioneer of effective online activism. At the beginning of President George W. Bush's first term, MoveOn.org was smaller than FreedomWorks currently is. But it grew so fast over the intervening years that by the 2004 election it had almost 700,000 donating members who gave an average financial contribution of $43.31 for a total of $32 million raised to defeat Republican candidates.

In the 2006 elections, progressive groups like EMILY's List and MoveOn.org worked together through the Democratic Party to take back control of Congress from the Republicans. There are now eighty-three Democratic members in the Congressional Progressive Caucus (CPC)—up from just six in 1991—making it the largest caucus within the Democratic Party in the U.S. Congress. That means the CPC has a big say in who runs the House. EMILY's List–sponsored Far Left Rep. Nancy Pelosi is the Speaker of the House. Of the twenty standing committees in the House of Representatives, half of them are chaired by CPC members. EMILY's List's progressive candidates also now hold governorships and cabinet positions. MoveOn.org knew it had a strong progressive candidate in Barack Obama and broke tradition by endorsing and campaigning hard for him in the Democratic primary over Hillary Clinton.

Founder Ellen Malcolm credits the PAC's powerful influence along with EMILY's List member Nancy Pelosi for the passage of health care reform. According to Malcolm, "In only twelve elections, you and I through EMILY's list have literally changed the face of power in America."

Without progressives like Nancy Pelosi and Barack Obama running Washington, and eighty-three progressive members in the CPC, the $786 billion government stimulus would not have passed. Neither would have government-run health care. We would not face trillion-dollar deficits. Even the Wall Street bailout would not have passed without the support of key progressives such as President Obama, Speaker Pelosi, and Rep. Barney Frank.

FIGHTING FIRE WITH FIRE

The lesson for fiscal conservatives is to get involved. Together, we can do the same for limited government as they've done for expanded government. By getting together with our friends, maybe in our basements like Malcolm and her friends, we can build large networks of individuals who each give a little time and money toward the goal of taking over the Republican Party. Together, it adds up to lots of time, lots of money, lots of influence, and, ultimately, a return to a government of, for, and by the people.

That's the only way this works—not from the top down, but from a broad network of like-minded individuals all doing a little more for their democracy through an existing party infrastructure. Having been to many town hall meetings, tea parties, rallies, protests, and marches over the past year, we know there are more than enough people on our side who can build a powerful network going into the 2010 fall election season and beyond.

The "progressives" worked hard and succeeded in enacting their big-government agenda. We can do the same with an equal commitment to good ideas and hard work and effective organization.

We should start by replacing those in the Republican establishment who don't agree with fiscal conservatives who are willing to cut spending and shrink government. Then we need to be there 365 days a year to keep them from joining the go-along-to-get-along club once they get to Washington—which goes along spending our money to get a long list of big-money contributors to their campaigns.

This is precisely where our movement has failed in the past. We've long discounted the value—and the work involved—in organizing taxpayers to protest out-of-control spending and the growth of government. "Our side doesn't do that," skeptics often said. "They have jobs and families." Instead, fiscal conservatives built world-class think tanks in the belief that our ideas were so powerful that simply being right would win the public policy battles.

The incentives are the opposite for the big-money contributors we're fighting. Washington lobbying is currently the best return on investment going, and not just because the stock market is in the tank. The federal government controls the spending of trillions of dollars every year. So big business, big labor, trial lawyers, and other well-heeled interests spend lavishly on lobbying, with the expectation that it will more than come back to them in the form of government handouts and sub-sidies.

Take the Wall Street bailout, for example. The Center for Responsive Politics reported in February 2009 that the "companies that have been awarded taxpay-ers' money from Congress's bailout bill spent $77 million on lobbying and $37 million on federal campaign contributions. . . . The return on investment: 258,449 percent." Sheila Krumholz, the center's executive director, commented, "Even in the best economic times, you won't find an investment with a greater payoff than what these companies have been getting. Some of the companies and industries that have received payments may now consider their contributions and lobbying to be the smartest investments they've made in years."

Without a doubt, this is exploitation of the many by the few.

Usually, Big Government and Big Business get away with this because the cost per taxpayer is too small for us to believe it is worth fighting over. Economists call this the "collective action" problem. It's all about concentrated benefits and dispersed costs. This idea was first spelled out by Mancur Olson, who said, "The taxpayers are a vast group with an obvious common interest, but in an important sense they have yet to obtain representation. The consumers are at least as numerous as any other group in the society, but they have no organization to countervail the power of organized monopolistic producers. . . . There are vast numbers who have a common interest in preventing inflation and depression, but they have no organization to express that interest."

So if the Tea Party movement wants to be politically effective in turning an ethos into public policy, we need to take over the Republican Party. By seizing control of the party, we can spend our time focused on ideas and use the party infrastructure that has been built over the past 156 years. And among the two parties the Republicans have at least at times been on the side of fiscal restraint and already have some of us in their ranks.

Notice that we call for a hostile takeover. We didn't say "join the Republican Party." We need to take it over. The commonsense values that define the Tea Party movement, like the belief that government should not spend money it does not have, puts us in the broad middle of American politics. That means the existing parties, if they covet the votes of this broad constituency, need to gravitate toward our values and our issues to get elected.

TESTING THE THEORY

Our strategy was put to the test soon enough. On September 29, 2009, New York governor David Patterson announced that a special election would be held on November 3 to fill a recently vacated U.S. House seat in the 23rd District. This was the same day that New Jersey and Virginia would elect their governors.

We, along with nearly a million activists who had joined together for the historic 9/12 March on Washington, were still smiling about what we had accomplished together. Hadn't this united demonstration of grassroots resolve sent a powerful message to Democrats and Republicans alike? Things now would change, wouldn't they?

New York law does not allow for primaries in special elections to fill vacant House seats. Instead, party leaders handpick the candidate. The Republican establishment—eleven committee chairmen in closed-door conclave—had anointed state assemblywoman Dede Scozzafava. Considered "electable," she was so left-leaning on core economic issues that she was virtually indistinguishable from her liberal Democratic opponent. Scozzafava supported Obama's government stimulus bill, the union-favored "card check" bill to end secret ballots in the workplace, and a government takeover of health care.

The Conservative Party of New York, which often backs Republicans, refused to endorse Scozzafava. Instead, it nominated Doug Hoffman, a fiscal conservative who had never run for political office. But the national Republicans, including Republican National Committee chairman Michael Steele, threw their support behind Scozzafava. The Republican Party establishment once again stood against grassroots citizens. A line was drawn in the sand.

As is typically the case at FreedomWorks, we first learned about this battle from the activists on the ground in New York. The political establishment had it wrong, they said. Scozzafava could not win; she was wrong on the core economic issues—the very issues that voters in the 23rd District cared about most. She would quite simply fail to compel voters looking for a real change from the deficit-spending agenda of the Pelosi Democrats in Washington.

Former Democratic House Speaker Tip O'Neill famously said that "all politics is local." He was giving advice to Democrats, of course. Democrats run on what they can bring home to the district, building a voting coalition of constituencies that want something from the federal government. Republican voters, on the other hand, tend to look for something else, a more national vision. They want fiscal responsibility, a good climate for economic growth and jobs, and to be left alone.

And so it went with this race. Voters were looking for a candidate who was willing to say *no* to new spending, someone who would take a principled stand for spending restraints and against more government meddling in our health care. We believed Doug Hoffman was that candidate; moreover, we believed he would be the winning candidate.

Dick got involved in the race early and endorsed Doug's thirdparty challenge. "My own view right now is the myth that you have to be a moderate—a Democrat lite—to win in the Northeast probably has less standing now than in any time since I've been in politics," he told the *New York Times*. "The small-government candidate in the Republican Party—or running as an independent—is going to be the one to draw the energy of these voters." Steve Forbes, FreedomWorks' vice chairman, and Sarah Palin endorsed Hoffman as well, making this the race that the national media was following, a measure of the new Tea Party.

As would become a pattern in elections to come, the Tea Party movement "nationalized" this election. Activists from all over the state of New York were getting involved in Doug Hoffman's underdog bid for Congress. "Word of his insurgent campaign spread locally and nationally, via online activist groups and publicity from conservative broadcaster Glenn Beck," the *Los Angeles Times* reported. "The campaign attracted volunteers like Jennifer Bernstone, a performing artist in Canastota, New York, who had never been involved in politics. Bernstone had been seething ever since President George W. Bush agreed to bail out teetering

Wall Street banks in late 2008. She snapped into political action a few weeks ago, after the GOP nominated
Scozzafava. 'Dede is more liberal than the Democrat,'
Bernstone said, as she put in a day of campaign work
that began at 8:00 A.M. and was likely to end after mid-
night." FreedomWorks staff and volunteers joined the
efforts, going door to door with voter educating materials and distributing literature outside polling places
up until the last moments of the campaign.

It was an amazing run to the finish line. In just two
months, Hoffman went from totally unknown and unfunded to the leader in virtually every poll of likely
voters. Siena College polling showed the dramatic rise
in support for the Conservative insurgent candidate,
who went from 16 percent in late September to a leading position of 41 percent support two days before the
election. Scozzafava, conversely, went from 35 percent
support in late September to 20 percent in late October.
She was in third place, trailing both the Conservative
and the Democrat.

The erstwhile Republican dropped out of the race
with just three days remaining, echoing the Left's message by blaming the "hate and lies and the deceitfulness" of her opponents for the dramatic collapse of her
campaign. In truth, Scozzafava had fallen so far behind
in the polls that she could not reasonably expect to win.

The voters had rejected her candidacy based on their concerns about a federal government that was out of control. She had not been able to make a credible case that she would go to Washington and help turn things back around.

This cleared the way for an easy Hoffman victory. The latest Siena College poll had Doug up by a five-point lead against the Democrat, Bill Owens, on November 1, just two days before the election. It was an incredible, meteoric rise from obscurity for the politically unpolished, but principled, fiscal conservative.

The nascent national grassroots movement built upon the founding principles of liberty had been ignored, then dismissed, and then ridiculed as a tempest in a teapot. Working together on a clear mission, we were all about to demonstrate the real world impact of good ideas combined with "boots on the ground." Together, we would finally demonstrate to the political class the true power of our decentralized, leaderless community.

We had waited and worked a long time for this, since the 2008 battle over TARP, when the colluded "crony capitalism" of Wall Street and official Washington was defeated by the people. That victory was far too temporary, snatched back by a desperate establishment within days of their first defeat on the floor of the U.S.

House of Representatives. We would take back some of that ground on November 3 in upstate New York.

And then Dede Scozzafava, the favored candidate of the Republicans, endorsed the Democrat, Bill Owens. "Since beginning my campaign, I have told you that this election is not about me; it's about the people of this district," she claimed. "It is in this spirit that I am writing to let you know I am supporting Bill Owens for Congress and urge you to do the same."

Owens ended up winning by capturing 48.3 percent of the vote, with the Conservative Hoffman receiving 46 percent. Everyone agreed that Scozzafava's decision to team with the Democratic Party had been the margin of difference in Hoffman's defeat.

The political bait and switch of a "Republican" to a Democrat was a bitter disappointment, but it was a stunt that we would see again from the establishment's candidates. The betrayal always arrived at the moment of personal defeat and disappointment, publicly rationalized as being in the best interests of "the people."

Did the Scozzafava debacle disprove the strategy of taking over the Republican Party? Absolutely not. It was the first in a multistep recovery for the GOP. The battle over the 23rd Congressional District in New York was the place where the Tea Party movement put the Republican establishment on notice: Ignore us at your peril. It was a necessary step on the path to recovery and political credibility. Now they knew we meant business. The Tea Party delivered and we were on our way to acceptance.

PARTY ANIMALS

Political parties, remember, are always intellectually and morally inferior to good ideas. Candidates of one party can sometimes just as easily run with the other party. They are hollow vessels that can remain conveniently empty of any semblance of a coherent worldview or guiding principles. Just ask Pennsylvania senator Arlen Specter, who fled the Republican Party at that very moment when it became clear that a true fiscal conservative, Pat Toomey, was about to take his job from him in the Republican primary. For a short time, Specter the defector was a Democrat. Now he is fully retired from politics by the voters of Pennsylvania, perhaps seeking the "End-of-Job Counseling" promised on the protest sign held up at the September 12 Taxpayer March on Washington. Who knows, the appropriation of funds for this program may be buried in the $786 billion government stimulus Specter voted for in 2009.

Beyond the 23rd Congressional District in upstate New York, there were other signs that the Republican establishment still did not understand the grassroots uprising that was going on across America.

The *New York Times* best summed up the strategic divide between the establishment and the grassroots ground troops that made up the Tea Party: "Ms. Scozzafava fit the model of candidate advocated by Republican leaders like Mr. Steele and Senator John Cornyn of Texas [the chairman of the National Republican Senatorial Committee]: one whose views might not be in keeping with much of the national party but are more reflective of the district."

"A primary is unfolding in Florida, where Gov. Charlie Crist, who is running for the Senate, is facing a challenge from a conservative, Marco Rubio, the former Florida House speaker. Mr. Crist has come under fire from conservatives for, among other things, supporting Mr. Obama on his economic stimulus package. . . . Mr. Cornyn said that he did not see other situations where Republicans could face a similarly divisive primary. He said he expected Mr. Crist to win the nomination."

The party illuminati—the perennial class of political operatives that represent the collective wisdom of Republican political strategy—continued to draw all of the wrong lessons from the 2006 and 2008 drubbings that had left Republicans with almost nothing. To win again, they counseled, the party needed to run candidates who acted like Democrats: Republicans should become Democrat-lite.

Wasn't this the complacent attitude we were fighting against in 1994? Wasn't this the exact strategy that had given Democrats a permanent, forty-year majority in the House of Representatives before the Contract with America?

Why do Republicans insist on acting like Democrats in hopes of regaining political power, while Democrats act like fiscal conservatives in order to win?

Somewhere along the way to "permanent majority" status, the Republicans who held Congress forgot or abandoned their national vision, letting parochial interests dominate their legislative decision-making process. The "Spirit of '94" was replaced by a far narrower, politically defined choice criteria. Their question, once articulated by a national vision for America, became: How do we hold our congressional seats and political power? The aberrant behavior, politically defined spending priorities, and scandals involving undue special interest influences that ended up defining the Republicans in 2006 were all a direct consequence of this shift in focus from policy to political power.

Nowhere was this turn more evident than in the dramatic collapse in fiscal discipline and the corruption of the budgeting process. A national vision of fiscal responsibility—"We will spend your money carefully and we will keep your taxes low"—was replaced with what Jack Abramoff infamously referred to as his "favor factory." Going into the 2006 elections, one Republican leader actually defended a highly dubious spending provision earmarked for the congressional district of an embattled incumbent, saying it was a reasonable price to pay for holding a Republican seat. What was most remarkable was not even the admission itself, but that he was so comfortable publicly acknowledging his motives.

Nancy Pelosi skillfully turned this Republican hubris on spending into a Democratic majority, promising to "drain the swamp" in Washington, D.C., and bring a new era of fiscal responsibility, transparency, and accountability to the appropriations process. She never did that, of course, but like candidate Barack Obama two years later, she and her fellow Democrats turned Republican failures into a politically potent strategy. Two election cycles later, the Democrats, dominated by their "progressive" wing, had supermajorities in both the House and the Senate and control of the White House.

In 2009 we still had our work cut out for us. But the winds of change were in the air. Opponents who seemed invincible just months ago were suddenly vulnerable. Activists in Massachusetts and Florida had watched events unfold in New York and realized that something special was happening. It wasn't just a protest anymore.

8

THE NEW CENTER OF
AMERICAN POLITICS

A lmost 150 years after the original band of tea partiers invaded Boston harbor, Massachusetts again
played host to an historic act of political independence.
In 1773 George Hewes and his fellow patriots stood up
for the right to determine their own fate as free men.
In 2009 American citizens stood on their forefathers'
shoulders and used that right to send a clear message to
the political establishment: it's still up to us. The power
is in the people's hands. Ignore it at your peril.

Working closely with independent chapter leaders and Tea Party organizers across the country,
FreedomWorks often holds the advantage of knowing
the facts on the ground long before the soothsayers in
Washington proclaim them to be true. The Scott Brown
phenomenon was no exception.

In December 2009 we were approached by several activists who told us the race in Massachusetts for the Senate seat left vacant when Ted Kennedy died was competitive. Conventional wisdom was wrong, they said. A practicing attorney and former Massachusetts National Guard JAG, Scott Brown held many of the principles we believe in and was swaying voters in a state that had grown weary of liberal rhetoric.

A LONG SHOT

Admittedly, we were doubtful. After all, this was Massachusetts. Obama had just received 62 percent of the vote in 2008 and the state hadn't elected a Republican senator since 1972. Moreover, the race was an explicit referendum on Obamacare and the Left's agenda more broadly. It was a referendum on big government—and it's not called "Taxachusetts" for nothing. If elected, Democratic candidate Martha Coakley promised to support the work of the president Massachusetts had overwhelmingly supported. Scott Brown, on the other hand, vowed to be the decisive forty-first vote in the Senate against policies that expand the size of government and add to the deficit.

Because of this, Tea Party activists and grass-roots conservatives saw the election as an opportunity to slow or even halt the agenda of the Left. Filling the seat previously held by liberal lion Ted Kennedy with a fiscal conservative would have been more than just a moral victory. It would have sent the forty-first Republican vote to Washington and shattered the Democrats' filibuster-proof majority in the Senate. Limited-government advocates realized: Scott Brown's election has national implications; we needed Brown to win. But could Brown actually win? The election was a little over a month away and most political commentators were writing it off as a shoe-in for the Democrats.

However, it has been our experience over the years that grassroots activists are often more knowledgeable than those whose job it is to understand the political climate. So we placed a phone call to a part-time employee working for us in Massachusetts named Matt Clemente. Clemente was a twenty-year-old student at the College of the Holy Cross in Worcester who was well connected with the Tea Party movement that was developing in Massachusetts. He had spoken at Tea Party events in both Worcester and Boston and was in close contact with leaders from across the state. If the Tea Partiers in Massachusetts were getting behind the Brown campaign, Clemente would know.

Matt told us that he had only seen one Martha Coakley sign in his hometown of Milford compared to dozens of Scott Brown signs. He told us that many local talk-radio hosts had gotten behind Brown and buzz about the race was dominating the airwaves. He said that almost everyone he talked to said that they were voting for Brown, even fellow college students who had passionately supported Obama's campaign only a year before. Although these were promising signs, Brown was still lagging in the polls, and we knew that without a strong effort on the part of the grassroots community he stood little chance in a state like Massachusetts.

Clemente told us that he would call back once he had spoken with fellow activists at a Worcester Tea Party meeting about the election. He was amazed by what he learned. The meeting was held in a small VFW hall that was filled beyond capacity. There weren't even enough chairs for everyone to sit down. More important, everyone was talking about their contributions to the Brown campaign. Some had been holding signs earlier in the day, others were making phone calls, and some were even taking time off from work to go door-to-door. The meeting that Clemente had expected to draw a few dozen people instead boasted a few hundred, and all of them believed that their next senator could very well be Scott Brown.

Tea Party activists from across the state (and ultimately from across the country) were drawn to Brown because they saw themselves in him. He campaigned like he was one of us. He was hard-working and dedicated. He was genuine. He didn't sit back and let consultants run the show. He shared our hopes and understood our concerns about the growth of government. Disproving the notion that Tea Partiers were extremists who existed outside of the mainstream, Brown's campaign demonstrated that the Tea Party movement represents the center of American politics.

After talking to Matt Clemente about the excitement within the grassroots community, we decided that it was time to get to work. We were going to do everything we could to help the in-state activists. We started by appointing Clemente to be our Massachusetts state director. He would be our point man, our eyes and ears on the front lines. Then we tasked FreedomWorks' campaign manager Nan Swift with creating a side-by-side comparison of the two candidates. We wanted a document to distribute to Massachusetts's voters that would tell them where each candidate stood on essential issues. With the election just about a month away, there was little time to waste.

As the Massachusetts Tea Party movement grew stronger by the day, we prepared to make our contribution. First-time activists had gained valuable experience in recent months, and as a result, they became more organized and better equipped to aid in the fight. Clemente told us countless stories of small-business owners or nurses or schoolteachers or even union members who he met with on a daily basis. Time and time again he would hear the same story of someone who had never been politically active but who was now working with the campaign. More and more people were fed up with what was going on in Washington, and there was a growing sentiment that it was time for them to do something. As these individuals began to unify, they became a massive grassroots network made up of members all working toward one goal: stopping the far left's big-government agenda.

AMERICA TAKES NOTICE

As a first-time activist, Mark Reeth—one of Matt Clemente's classmates at Holy Cross who helped him organize the door-to-door effort for FreedomWorks—was astonished by the dedication of the local Tea Party members. "It was so inspiring," Mark said. "I mean, when you have parents asking neighbors to watch their kids so they can spend a few hours going door-to-door, you know that you're involved with something pretty amazing."

The support for Brown was overwhelming. And as election Tuesday drew closer, America began to take notice. It wasn't long before our office was flooded with phone calls and e-mails from people across the country who wanted to help. Tea Party activists from every corner of America were ready to join in the fight. People were making the drive from Connecticut, Rhode Island, New Hampshire, Maine, even as far as Florida and Alabama, to help with the cause.

One activist who made the trek from Alabama was a thirty-four-year-old Marine Corps veteran, dad, and entrepreneur named Rick Barber. Rick told us he "went to Massachusetts in the hopes of killing the health care bill"—something he had spent much of 2009 fighting. In an appearance on *Good Morning America*, Rick captured his motivation in six words: "In 2010 all politics are national."

Activists across the country agreed and support for Brown poured in from all over the country. We heard from Matt Clemente that hotels were booked solid, phone banks were pumping out thousands of calls a day, and you couldn't drive for more than a minute without seeing a SCOTT BROWN UNITED STATES SENATE lawn sign. Something important had begun to happen; something that few political observers outside of the Tea Party movement could have anticipated. Suddenly the poll numbers began to shift. With every new poll, Brown gained a few points. First he was within 15; then he was within 10, then 5; then he was within the margin of error.

ON THE BRINK

But it wasn't all good news coming out of Massachusetts at that time. One week before the election, an in-state activist forwarded an e-mail to FreedomWorks entitled WARNING TO TEA PARTY ACTIVISTS: DON'T EVEN THINK ABOUT VOTING FOR SCOTT BROWN! The original e-mail was from Carla Howell and Michael Cloud, two prominent small-government advocates. Howell and Cloud are well-respected members of the Massachusetts Tea Party movement, and the e-mail was sent to grassroots conservatives and libertarians statewide. They have both run for statewide office and have repeatedly—and almost successfully—led the charge to repeal Massachusetts's state income tax.

In part, their e-mail read, "The Republican Party . . . [is] trying to scare and stampede you and us—Tea Party activists, Town Hall Meeting protesters, and tax cutters—into closing our eyes, holding our noses, and voting for Brown—out of fear that the alternative is even worse. They are wrong. . . . You have a radically better choice. A choice that will advance the Tea Party Cause. A choice that will give us REAL Tea Party candidates and allies in November."

This, of course, is the constant tension in politics— deciding whether or not to let the unelectable "perfect" be the enemy of the electable "good." Unfortunately, as long as free speech is limited in the United States by unconstitutional campaign finance restrictions, it will remain a two-party country. This, of course, is why bipartisan campaign finance laws have been so popular with incumbents.

The choice of which Howell and Cloud wrote was a third-party libertarian candidate named Joseph Kennedy. By all measures, Kennedy's views on the free market system and the role of government are in line with most Tea Party activists. He was against the TARP Wall Street bailout, against other government bailouts, against the health care bill being rammed through Congress, against the cap-and-trade energy tax, and in favor of individual liberty. The only problem was that polling showed he was pulling around 1 percent of the vote. As if that wasn't enough, the market had clearly spoken: individuals had shown their support by giving Brown more than $15 million in campaign contributions, while Kennedy had attracted just $18,000. If Kennedy was slightly better on the issues, we disagreed with Howell and Cloud's assessment of Scott Brown. A losing candidate who may be slightly better on the issues is not a "radically better choice" than a winning candidate who will be with you on the biggest legislative fights of the day.

The movement realized this and saw it was in our best interest to support Brown. The inescapable truth was that supporting him was the right decision in order to prevent the election of Martha Coakley as the critical sixtieth Democratic senator and the passage of every bill on the liberal wish list. The Tea Party activists in Massachusetts agreed. Maybe we're not all as naive as the Left and the media hope and say we are.

This last-minute division from Howell and Cloud proved inconsequential as most activists realized what was at stake and accepted the current reality of our system. Brad Wyatt, a member of the Worcester Tea Party who let the Brown campaign set up shop in his office building, saw his vote for Brown as a once-in-a-lifetime opportunity to affect the course of the nation. "Many of the Tea Party citizens preferred the libertarian Joe Kennedy (and I agreed with his ideological stances as well), but Scott Brown had the best chance to win against Martha Coakley and the Democratic machine in Massachusetts, and so the Tea Party people solidified their support behind Scott."

So the libertarians may generally be counted among the Tea Partiers, but does that just confirm the nagging accusation that our fiscal conservatism puts us on the fringe? Which way Massachusetts's independent voters turned would let us know—and they make up more than 50 percent of the state's electorate, so they would also decide the election. Would they line up with the Tea Party movement behind the same candidate?

Mary Anne Pappas, the grandmother of Freedom-Works' vice president for public policy Max Pappas, proved to be a harbinger of things to come. A long-time independent, Mary Anne has been casting votes in Massachusetts since she marked a ballot for Franklin Delano Roosevelt in 1944. She's been voting in Massachusetts long enough to have voted for Ted Kennedy "several times," and she had voted for Martha Coakley in the Democratic primary just weeks earlier. But by no means has she been a party-line voter. She told us, "I like to know the records of the people I'm voting for and ask around my circle of friends for input. I have also voted Republican at times." As independent voters like Mary Anne go, so go most elections.

Max spoke with his grandmother while home over Christmas about the race. He pointed out that Brown would be the forty-first vote needed in the Senate to sustain a filibuster against Obamacare. Mary Anne responded, "I'm not too happy with the bill anyway, and if this will help, I'll vote Republican and suggest to my friends that they do the same."

The election was drawing closer, and it was becoming evident that momentum was on our side. Matt Clemente and the other FreedomWorks activists were pounding the pavement day in and day out to get our side-by-side candidate comparison into the hands of as many likely voters as possible. In-state Tea Party activists were making themselves available to do anything and everything that the campaign asked of them. Scott Brown went from being a relatively unknown state legislator to the champion of the grassroots Tea Party conservative movement overnight.

Coming into work on Election Day, we noticed that something about Washington felt different. There was an undeniable excitement in the air. After proponents of smaller government had suffered major defeats with TARP, the bailouts, and the stimulus bill, we knew that we finally had an opportunity to send a message: No more. No more frivolous spending and bureaucratic waste. No more lack of transparency and backroom deals. No more failed promises. No more tax increases. No more assaults on our freedom. But it wasn't up to us to send that message. It was up to the voters of

Massachusetts. All that we could do was sit and wait for the results to come in.

Meanwhile, in Massachusetts, the final push was on. Clemente sent e-mails and text messages to everyone he knew reminding them of the importance of their vote. Every town in the area had Brown supporters waving signs and directing people to the polling places. After months of tireless work, activists pushed on, making phone calls, walking door-to-door, and driving those who were unable to drive themselves to the ballot box. It is easy to grow restless over the course of a long election cycle. It is easy to become complacent and assume that victory is at hand, but the Massachusetts grassroots community did not give in to such temptations. From the minute they became involved with the campaign until the final vote was counted, the Tea Partiers gave Scott Brown everything that they had.

The impossible happened as a result. The state of Massachusetts rejected big-government liberalism and voted a Republican to the United States Senate: a Republican who explicitly campaigned on being the decisive vote against the massive health care overhaul so closely tied to Ted Kennedy. And former Kennedy voters were making this happen.

That night, as the FreedomWorks D.C. staff gathered at a local bar to watch the vote results come in, we received an e-mail to our cell phones. It was from Matt Clemente and the subject read, YOU'RE WELCOME. The message simply stated, FROM ALL YOUR FRIENDS IN MASS.

We had done it.

The Tea Party community had come together around an electable candidate and had worked vigorously to get him into office. We clearly were now a true political force with the power to affect elections and influence policy. On January 18, we had gone to sleep in a country in which big-government politicians and Washington elites shouted over the voices of the American public. On January 19, we awoke to a revitalized America. We had awoken to a country in which the voice of the people still mattered, a country in which a group of individuals united by a common cause could still make a difference, a country in which anything was possible. The media and the powerful in Washington had no choice but to take notice.

After the election was over, Matt Clemente told us about his eighty-seven-year-old Aunt Ginny. She was a lifelong Democrat. In her first presidential election she, like Max's grandmother Mary Anne, voted for FDR, and she, like many Massachusetts residents, adored the late Ted Kennedy. She had never voted for a Republican and Matt was under the impression that she never would. But in Scott Brown she saw more than just another Republican candidate. She saw the same thing that led grassroots activists to drive cross-country and the same thing that led Tea Partiers to dedicate months of their lives to one man's campaign. In him she saw a much needed check on one-party domination and the corrupting influence such power can have on even those with the best intentions. In him she saw the future of America, the future of freedom, and a return to sanity in Washington. That is why she, like 52 percent of Massachusetts voters, cast her ballot for Scott Brown.

To earn the grassroots support, Republicans will have to be bold on policy that will get to the heart of the problem: Americans think government has grown too big and spends too much. Our job as the voters in this country is to supply the boldness for party leaders by making it clear we'll be participating in November for those who are as bold as we are in our desire to limit Washington's power.

That is how we exert our influence over the Republican Party.

A CONTRACT *FROM* AMERICA

Grassroots activists have already presented great ideas we will push aspiring politicians to embrace—for their own good, and for the country's sake. One of the best ideas comes from Ryan Hecker, a Tea Party leader in Houston. Ryan realized the Republicans, having so recently failed to govern as fiscal conservatives, didn't have the credibility to present their own Washington-created contract to the American people. So he set out to build the infrastructure that would enable the American people to offer one themselves for politicians to sign on to. Ryan launched a Web site in mid-2009 called ContractFromAmerica.com to gather ideas from across the country for a new contract, a Contract from America.

Over several months he received thousands of ideas from every corner of the country, boiled them down to twenty-two by combining the most commonly sub-mitted points into coherent policy positions, and then put the list out for a national online vote. He received hundreds of thousands of votes from grassroots activ-ists on his site and presented the final ten as the Con-tract from America before forty thousand activists at our Tax Day Tea Party held on April 15, 2010, in front of the Washington Monument.

THE CONTRACT FROM AMERICA

We, the undersigned, call upon those seeking to re-present us in public office to sign the Contract *from*

America and by doing so commit to support each of its agenda items, work to bring each agenda item to a vote during the first year, and pledge to advocate on behalf of individual liberty, limited government, and economic freedom.

Our moral, political, and economic liberties are inherent, not granted by our government. It is essential to the practice of these liberties that we be free from restriction over our peaceful political expression and free from excessive control over our economic choices.

The purpose of our government is to exercise only those limited powers that have been relinquished to it by the people, chief among these being the protection of our liberties by administering justice and ensuring our safety from threats arising inside or outside our country's sovereign borders. When our government ventures beyond these functions and attempts to increase its power over the marketplace and the economic decisions of individuals, our liberties are diminished and the probability of corruption, internal strife, economic depression, and poverty increases.

The most powerful, proven instrument of material and social progress is the free market. The market economy, driven by the accumulated expressions of individual economic choices, is the only economic system that preserves and enhances individual liberty. Any other economic system, regardless of its intended pragmatic benefits, undermines our fundamental rights as free people.

PROTECT THE CONSTITUTION. Require each bill to identify the specific provision of the Constitution that gives Congress the power to do what the bill does.

REJECT CAP & TRADE. Stop costly new regulations that would increase unemployment, raise consumer prices, and weaken the nation's global competitiveness with virtually no impact on global temperatures.

DEMAND A BALANCED BUDGET. Begin the constitutional amendment process to require a balanced budget with a two-thirds majority needed for any tax hike.

ENACT FUNDAMENTAL TAX REFORM. Adopt a simple and fair single-rate tax system by scrapping the internal revenue code and replacing it with one that is no longer than 4,543 words—the length of the original Constitution.

RESTORE FISCAL RESPONSIBILITY & CONSTITUTIONALLY LIMITED GOVERNMENT IN WASHINGTON. Create a blue-ribbon task force that engages in a complete audit of federal agencies and programs, assessing their constitutionality, and identifying duplication, waste, ineffectiveness, and agencies and programs better left for the states or local authorities, or ripe for wholesale reform or elimination due to our efforts to restore limited government consistent with the U.S. Constitution's meaning.

END RUNAWAY GOVERNMENT SPENDING. Impose a statutory cap limiting the annual growth in total federal spending to the sum of the inflation rate plus the percentage of population growth.

DEFUND, REPEAL, & REPLACE GOVERNMENT-RUN HEALTH CARE. Defund, repeal, and replace the recently passed government-run health care with a system that actually makes health care and insurance more affordable by enabling a competitive, open, and transparent free market health care and health insurance system that isn't restricted by state boundaries.

PASS AN "ALL-OF-THE-ABOVE" ENERGY POLICY. Authorize the exploration of proven energy reserves to reduce our dependence on foreign energy sources from unstable countries and reduce regulatory barriers to all other forms of energy creation, lowering prices and creating competition and jobs.

STOP THE PORK. Place a moratorium on all earmarks until the budget is balanced, and then require a two-thirds majority to pass any earmark.

STOP THE TAX HIKES. Permanently repeal all tax hikes, including those to the income, capital gains, and death taxes, currently scheduled to begin in 2011.

One of the primary concerns we repeatedly hear at rallies is that no one in Washington is listening. The Democrats increased this perception by passing the health care legislation, employing extraordinary procedural trickery, that much of the country had vociferously opposed. Instead, they belittled the protesters. If the Republican Party ignores this Contract from America, they will be signaling to us that they aren't listening, either. They know it exists—we've told many of the leaders in Washington personally. And some have wisely embraced it. It is our job to make sure many more do.

If the Republican Party wants to find its way back into power this year, it will need to live up to its promises of limited government, embrace the Tea Party movement, and allow the movement to guide the agenda. This is the formula for success for everyone who believes in limited government, whether they currently belong to the tea parties or the Republican Party or the Democratic Party or some other party or no party at all.

Republicans want to get elected, and we can influence the outcomes of their elections. As fiscal conservatives, we are the center of American politics. We are the elusive independents in the middle who decide elections year after year. So they need us.

But we need them—and their well-built infrastructure—too.

The establishment and the media will never understand that the goal of our peaceful revolution is not just getting Republicans elected. We should not waste our time being concerned when they make that simple accusation. We know our goal is to use the Republican Party to shift the center of gravity in Washington to small-government conservatives—that's to say, we aim to get Washington's agenda in balance with the convictions of Americans outside the Beltway.

NO TIME TO CELEBRATE

A seismic shift in American politics was being generated from the ground up.

The warnings came in waves. First there was the unexpected collapse of the initial TARP vote on the floor of the House of Representatives. And then came an explosion of opposition to the Democrats' $786 billion "stimulus" bill that no one had bothered to read before it was rammed through Congress. Then August saw the visceral opposition to a hasty attempt by President Obama to have government takeover of our health care. There was the special election in New York District 23, where a no-name citizen came within an inch of defeating the handpicked professionals of both political parties. Then Scott Brown's election sent tremors across the country.

The very foundations of a deeply entrenched establishment were starting to shake, but Republicans and Democrats alike chose to ignore what was happening. The telltale signs were there all along, but for some it was easier to pretend everything was going to be just fine.

So when this political earthquake finally hit Florida, Governor Charlie Crist had no idea it was coming. Crist was lauded by many in the GOP establishment as a new breed of Republican, a fiscal "moderate" they embraced as the answer to the drubbing of the 2008 election. For fiscal conservatives, he was an awful choice to send to the U.S. Senate. He seemed like the ultimate opportunist, easily morphing into the champion of the latest political fad.

As governor, he embraced "cap and trade" restrictions on energy use that put Floridians at an economic disadvantage without any tangible benefit to environmental quality. He had also essentially taken over the market for property and casualty insurance in the state, setting up a taxpayer-funded system that was a fiscal time bomb, not unlike Fannie Mae and Freddie Mac, set to go off with the next serious hurricane. And he had been dead wrong to embrace President Obama's waste-filled stimulus plan, the very one that had drawn the ire of Mary Rakovich and her small band of protesters back in February 2009.

In spite of all this bad economic judgment, when Crist announced his bid for the U.S. Senate, the Republican Party broke with the tradition of staying out of primary races and immediately endorsed Charlie Crist as the favored candidate.

Their thinking was superficial at best. Crist was popular. He had a massive political war chest of almost $10 million. And he was, they claimed, the future of the Republican Party: more moderate, more like the Democrats who had just swept the 2008 elections. Liberal columnist E. J. Dionne summed up the Republicans calculations best, writing two days after the announcement that "Florida will be one of the clearest tests of whether rank-and-file Republican voters are more interested in doctrinal purity, or in winning—even if it means nominating an Obama hugger." (Crist famously had been photographed hugging President Obama onstage; the image soon became a ubiquitous reminder of his support for the president's stimulus plan.)

Marco Rubio had a different perspective; he thought that a Republican should stand for limited government.

We first noticed Rubio through the eyes of Freedom-Works activists on the ground in the state of Florida. They saw the young Cuban American as a rare combination of principles and political talent, someone who was taking the ideas of liberty and driving the legislative agenda in the state capital in Tallahassee.

In 2008 FreedomWorks asked activists from across the country to nominate a "Legislative Entrepreneur of the Year" for leadership at the state level. Our man in Florida Tom Gaitens quickly nominated Marco for his progrowth, proconsumer votes and strong support for lower taxes and more freedom.

During his time as speaker of the Florida House of Representatives, Rubio supported the efforts to protect citizens against the repercussions of the Supreme Court's wrongheaded Kelo decision on eminent domain. Our team in Florida helped to champion the effort to protect the property rights of the residents of Rivera Beach and ultimately support provisions that would benefit all Floridians. Speaker Rubio was strong in these efforts and built a good working relationship with the boots-on-the-ground activists. These activists would rally in Tallahassee each session calling for lower taxes, less government, and more freedom. Each time, the speaker would greet them warmly with an open door.

THE RIGHT STUFF

Marco was already running for the U.S. Senate before Crist announced on May 12, 2009. He had no money and virtually no name recognition. Outside of his home turf in the Miami area and to a dedicated core of FreedomWorks activists, few really knew about him. It looked like an impossible uphill climb for the young challenger. How could you beat the sitting governor with huge name-recognition numbers, a pile of money in the bank, and the support of the entire political establishment from Tallahassee to Washington?

What he did have was a fidelity to good ideas, an ability to get things done, and a record worth supporting. This was exactly the kind of candidate newly minted citizen activists like Mary Rakovich was looking for. From day one, the early Tea Party movement started to spread the word on the ground, evangelizing about this fresh face who offered something different, something worth fighting for. They also had not forgotten how Crist embraced the president's stimulus and knew what that meant for his record on spending if he was elected to the Senate.

The activists attended local Republican meetings to recruit new support. Our folks on the ground were certain that Rubio would be competitive if he could get the attention of the voters. Rubio became a welcome guest at Tea Party events across the state, speaking at an April 2009 event in south Florida organized by Tom Gaitens and again in Orlando in July.

At their urging, Dick Armey endorsed Rubio early in the primary on July 14, 2009, in the hopes of shedding more light on his record that stood so tall against the political opportunism of the sitting governor. In a press release, Dick described Rubio as "an inspiring leader for the next generation of the conservative movement. His track record and conservative convictions are a breath of fresh air in a party looking for new leaders to advance the principles of limited government, lower taxes and economic liberty."

Yet despite the obvious advantages these fiscal conservatives found in Rubio, Crist enjoyed an enormous lead in the first Quinnipiac Polls. In late July, rumors were swirling in the Florida press of Rubio dropping out of the senate race and running for attorney general.

But the Tea Party had a champion and got to work. Grassroots leaders from around the state began inviting Rubio to speak at rallies, bolstering his grassroots exposure. In September, Dick attended a Rubio fund-raiser in Dallas and learned that the young lawmaker was not just a champion for Florida, but for small-government conservatives across the nation.

A few weeks later, Dick came to South Florida and held a rally in Coral Gables. Slowly, the Tea Party movement's support helped bring Rubio to the public eye. As the Tea Party continued to organize and grow, so did Rubio's popularity. By October the hard work had begun to pay off and Rubio was even in the polls. By December, he was in the lead.

On February 10, 2010, FreedomWorks PAC hosted a reunion of sorts. To mark the anniversary of the Crist–Obama hug, Mary Rakovich was joined in Fort Myers by 1,500 activists from across Florida on the same day and at the same location of Mary's first protest. The keynote speaker that evening: Marco Rubio. Activists raised more than $750,000 for Rubio's campaign, proving that grass roots can equal the establishment's fund-raising efforts.

Rubio's surge in popularity via the Tea Party movement led to a decision by Governor Crist to drop out of the GOP and run as an independent. The insider-anointed winner was so far behind he didn't want to risk a crushing primary defeat.

HERE TO STAY

In Massachusetts, activists worked overtime to win at the finish line. In Florida, they pounced sooner and altered the primary process. A similar story played out in Utah, where Senator Robert Bennett, a three-term Republican mainstay, was ousted at the state Republican convention by Tea Party activists. When the dust settled, a whopping 2,200 of the 3,600 delegates had been personally contacted by FreedomWorks staffers and local volunteers. The pro-freedom contingent even set up a booth on the convention floor, debating the opposition in person and winning votes up to the last moment.

Bennett was widely considered to be a "good guy" who was mostly reliable on Republican issues. But he was unreliable when it came to fiscal conservatism. Bennett was an ardent defender of appropriators' earmarking habits, and had sponsored legislation proposing an individual mandate for health insurance that became the basis of Obamacare. Most notably for the delegates from Utah, he had voted for the Wall Street bailout. As Bennett spoke to the gathering, the chant of "TARP, TARP, TARP" echoed across Convention Hall. Bennett was ultimately replaced by Tea Party underdog candidate Mike Lee, a staunch supporter of limited government and the very first signer of the Contract from America.

On *Meet the Press*, *New York Times* columnist David Brooks fumed, "It is a damn outrage." E. J. Dionne of the *Washington Post* wailed, "It's almost a nonviolent coup." Get used to it, guys.

Nothing about elected office should be comfortable. No one should be reelected for hazy memories of partial support and tangential involvement on important issues. With his defeat, Bennett signaled the expanding power of a movement that no longer bites its fingernails hoping for a miracle on Election Day. And as citizens and patriots, we should all be proud of that.

WE ARE A MOVEMENT OF IDEAS, NOT LEADERS

The Tea Party's call to arms that first began with Mary Rakovich in Fort Myers has evolved into one of the most potent political forces in American history. This citizen movement is so effective because it has been in large part self-organizing. The many branches of the Tea Party movement have created a virtual marketplace for new ideas, effective innovations, and cutting-edge tactics. We agree on the first principles of individual freedom, free markets, and constitutionally constrained government, but when it comes to how to best advocate these ideas, best practices come from the ground up, around kitchen tables, from Facebook friends, at Tuesday book clubs, or on Twitter feeds.

That's why the Tea Party ethos gives the political establishment—Left and Right—such uncontrollable fits. They don't know what to make of it. They don't know what to call it. They want to talk to the man in charge.

If they knew who was in charge, they could attack him or her. They could crush the inconvenient dissent of the Tea Party. Remember "the thirteenth rule" of Chicago street politics according to Saul Alinsky: "Pick the target, freeze it, personalize it, and polarize it." This works when your target is a hapless CEO about to be turned on the class warfare spit. It's proving much harder to demonize millions of patriotic citizens, mothers and daughters, fathers and sons, grandparents fearful that their great-grandchildren will never live the American dream.

The Tea Party is the product of a perfect storm of (1) broken Republican commitments, (2) the aggressive left-wing agenda of a Democratic regime motivated by redistributionist values that are antithetical to the values of most Americans, and (3) technological innovations that allow people to find one another, organize, and get essential information in real time from competitive sources.

We call this complex and diverse movement "beautiful chaos." Or better yet, to borrow Nobel Prize–winning economist F. A. Hayek's weighty notion: "spontaneous order." By this we reference what is now the dominant understanding in organizational management theory: decentralization of personal knowledge is the best way to maximize the contributions of people, their talents, and the total productivity of any enterprise, no matter how big. Let the "leaders" be the regional activists who have the best knowledge of the local personalities and issues. In the real world, this is common sense. In Washington, D.C., this is known as radical. Even dangerous.

Hayek argues that it is impossible to replace the decentralized wisdom of freely acting individuals with a top-down government-imposed hierarchy without destroying the knowledge needed to rationally allocate resources in society:

> Which of these systems is likely to be more efficient depends mainly on the question under which of them we can expect that fuller use will be made of the existing knowledge. This, in turn, depends on whether we are more likely to succeed in putting at the disposal of a single central authority all the knowledge which ought to be used but which is initially dispersed among many different individuals, or in conveying to the individuals such additional knowledge as they need in order to enable them to dovetail their plans with those of others.

Hayek, along with his mentor, Ludwig von Mises, used this essential point to explain why markets work and why socialism fails. Discounted by their peers in academia at the time, their predictions came, quite tragically, to pass. Much later, Hayek would brand the arrogance of big-government planners a "fatal conceit."

When you think about it, a decentralized model for social change is most consistent with the values of independence, self-reliance, and personal liberty that embody America. Those activists who gathered at Boston's Old South Meeting House in 1773 knew it. Thomas Jefferson understood this when he wrote, "I would rather be exposed to the inconveniences attending too much liberty than to those attending too small a degree of it."

SUPER-SIZED

The big-government crowd, on the other hand, is naturally drawn to the compulsion demanded by a centralized authority. They can't imagine an undirected social order. *Someone needs to be in charge.* "We can't give people a choice or they might take it," said Senator Ted Kennedy during a closed-door House–Senate conference committee dealing with health savings accounts.

Big government is audacious. It is conceited. It knows better. Government is, by definition, the means by which you are compelled by force to do that which you would not do voluntarily. Like pay high taxes. Or "purchase," by federal mandate, a government-defined health insurance plan that you cannot afford, do not need, or simply do not want. For the left, and for today's monolithically liberal Democratic Party, every solution to every perceived problem involves more government: top-down dictates from new laws enforced by new bureaucrats who are presumed to care more and, most important, know better what you need. "I'm from the government, and I'm going to help you whether you want it or not."

Would anyone voluntarily bail out strangers living thousands of miles away who lied on their applications to buy a home? Of course not! It's a stupid idea that rewards bad behavior. YOU CAN'T FIX STUPID, BUT YOU CAN VOTE IT OUT OF OFFICE, reads a popular Tea Party protest sign. Consumers in free markets uncorrupted by regulatory favoritism vote untold millions of times a day, punishing irrational behavior, bad actors, and liar loans with equal and swift justice. Government, on the other hand, socializes bad behavior, taking from the responsible and giving to the irresponsible.

That's not how tea partiers roll.

And that's not how most Americans roll, either. Matt Kibbe got a lot of heat from liberals for telling the *New York Times* that "Americans are just genetically opposed to socialism." But it seems like an obvious statement of fact. America is different. We are special because our founding was conceived in liberty. It was in the genes of the Sons of Liberty who risked their lives, fortunes, and sacred honor for an idea. That genetic code makes our family, our community, and our country unique from all the rest of the world. Dick remembers a conversation with a friend who had emigrated from Ethiopia and was so proud that he had just completed his naturalization requirements for U.S. citizenship. He couldn't stop talking about it. He spoke about the U.S. Constitution, about the founders, and about freedom. As proud as he was of his heritage, he was a different man: he was an American.

Those liberals now in control of our government seem bent on apologizing for the United States, striving to, in the words of Barack Obama, "remake America." They want to remake us to look more like European social democracies. Liberals don't talk about democratic socialism anymore; they prattle on about "social justice." They misuse the phrase. Justice means treating every individual with respect and decency and exactly the same as anyone else is treated under the laws of the land. As best we can tell, "social justice" translates to really wise elected officials (you know, smarter than you) redistributing your hard-earned income to their favored social agendas, all dutifully administered by a well-intentioned bureaucrat. In Europe, this translates into bloated social welfare programs that punish work; massive tax burdens, particularly on the working class through hidden value-added taxes that crush economic expansion; and structural barriers to opportunity for younger generations of have-nots trying to enter the workforce.

The politics of greed is always wrapped in the language of love. When you hear someone go on about social justice, read between the lines. More government control of health care is not really about improving access to health care; it's about controlling your health care.

If you want to comprehend the energy and passion behind the citizen activists who are fighting this corrosive ideology of redistribution, understand this: we believe that America's founders got it right and that Europe got it wrong. America is different because we are all about the individual over the collective. No one says it better than Howard Roark in his speech to the jury in Ayn Rand's *The Fountainhead*:

Our country, the noblest country in the history of men, was based on the principle of individualism, the principle of man's "inalienable rights." It was a country where a man was free to seek his own happiness, to gain and produce; not to give up and renounce; to prosper, not to starve; to achieve, not to plunder; to hold as his highest possession a sense of his personal value, and as his highest virtue his self-respect.

Individualism is the unity of purpose that binds the Tea Party movement into a cohesive community. And left-wingers, bless their hearts, will never get it. What looks like chaos to its detractors is the essential order to the Tea Party. Freedom unleashed is a potent force for social change.

THE ROAD AHEAD

The early, targeted efforts of concerned citizens like Mary Rakovich have now taken root, grown, and blossomed into a social phenomenon that is so powerful because it is not directed by any one mind, political party, or parochial agenda. The criteria for membership in the Tea Party is straightforward: stay true to principle even when it proves inconvenient. Be assertive but respectful. Add value and don't take credit for other people's work. Our community is built on the Trader Principle: we associate by mutual consent, to further our mutually shared goals of restoring fiscal responsibility and constitutionally limited government.

These were the principles that enabled the September 12 Taxpayer March on Washington to become one of the largest political protests in the history of our nation's capital. How do you get a million people to travel to Washington, D.C., from the four corners of the country, on their own dime, to join in a common purpose? In a word: freedom. It was a glorious day. It was fun. It was irreverent. The 9/12 Taxpayer March on Washington was created by a beautiful mob of peaceful citizens engaged in patriotic dissent. No one asked permission, and no one considered the possibility that it was not their born right to peaceably assemble. Is that what the political establishment hates so much about our community?

It is time to take America back. We need to reclaim America from the advocates of big government in both political parties, from the rent-seeking corporations eager to use the power of government to enrich themselves at the expense of consumers and taxpayers, and from the web of left-wing special interests who feed at the public trough and consider it their right to do so.

The political potential of the broad grassroots movement against big government that we are witnessing today should not be underestimated. There is a small-"l" libertarian, commonsense fiscal conservatism out there that transcends partisan definitions. These are independent voters who are united around the idea that government is spending too much money it does not have, and that government is getting involved in things, like controlling health care and running car companies, that it cannot do effectively, and should not try to do at all.

These highly motivated concerns about fiscal issues now represent the very center of electoral opinion among Republicans, most independents, and a growing number of Democrats who have developed buyer's remorse. Today the liberals who control Congress make even Bill Clinton look conservative by comparison and they are scaring Americans with their fiscal lasciviousness. This overreach is the stage upon which to build a revolt. We can take America back from moneyed special interests, leftist advocacy groups, and arrogant politicians. We can stop the monumental legislative threats to our economic liberties. Most important, perhaps, we can do these things by building a national community of activists—organized on the ground and connected online—that will be able to hold the next generation of political leaders, whether they are Democrats or Republicans, accountable for their actions.

The Tea Party is different. Consider the comparable events that led to the political backlash in 1994. That was a voter uprising that too quickly waned when the imminent threat of one-party rule under the Democratic establishment seemed contained. The new activists who had risen up to throw the bums out of power eventually left the playing field again, leaving our political system in the hands of politicians. Left unattended, these politicians, as they all eventually will, returned to tending to their own self-indulgent needs. At best, they became inconvenience-minimizers, eager to compromise for the lesser of two evils. At worst, they grew their own power at the expense of the American people and the fiscal health of our economy. All of the corruptions that followed—the ballooning federal debt, the frenzied spending, the political favors, the bailouts, and the government takeovers—now confront our economy, our futures, and the American way of life built upon freedom, opportunity, and prosperity.

This political boom-and-bust cycle, not unlike the government-generated business cycle that caused the housing bubble and the massive mistakes that went with it, generates periods of accountability followed by years of neglect and an inevitable slip back to business as usual. The problem with this cycle, beyond the policy damage done, is the difficulty in reversing the trend toward more government spending and more government control over our lives. With each new government program, the baseline of total spending is raised, phony budget estimates become very real red ink, and the federal take grows as a percentage of the total private economy. When a constitutional barrier is breached, as happened with the extraordinary ceding of power to an unelected secretary of the treasury under TARP, there is permanent damage done to that constitutional wall that stands between free citizens and a tyrannical government. When informal constraints against hasty legislative actions are torn, as happened with the Democrats' decision to create a massive new health care entitlement through parliamentary chicanery, there is no going back to the way it was before. A future Congress will certainly try to use its new power to enact sweeping legislation with similar tactics, permanently end-running the "cooling off" function the authors of the Constitution envisioned in their design of a deliberative Congress.

The Tea Party movement is rising up because we know we cannot leave public policy to the politicians, or to the "experts," or to someone else with a parochial agenda, a concentrated benefit that comes first, before the public good, and at your expense. The broad community of patriotic citizens that have stood up to take their country back from an unholy alliance of government power and privileged interests are making a difference in ways that defy easy comparisons to the boom and bust of other recent shifts in the political winds. The Tea Party has evolved from political revolt to social movement. We the people are that force more powerful, a force that can save our great nation for future generations.

The establishment doesn't like it one bit. They will kick and scream and throw every possible roadblock in our path.

But we suspect George Washington would love it. He, after all, demanded as much of us. "The preservation of the sacred fire of liberty, and the destiny of the Republican model of government are justly considered as deeply, perhaps as finally staked, on the experiment entrusted to the hands of the American people."

Or, as we like to say, freedom works.

EPILOGUE: CHANGING THE CULTURE IN TACOMA, WASHINGTON

I don't expect politicians to solve anybody's problems. . . . We've got to take the world by the horns and solve our own problems. The world owes us nothing, each and every one of us, the world owes us not one single thing. Politicians or whoever.

—BOB DYLAN

Every two years, like clockwork, a fresh crop of candidates from both political parties promise the voters that they will come to the nation's capital and act differently. They say they will spend taxpayer dollars more prudently; they will drain the swamp along the Potomac of the undue influence of well-heeled special interests; they will bring with them a new era of nonpartisanship, or civility, or transparency, or whatever their pollster says "moves the dial." On cue, each and every one of them promises to change the culture in Washington, D.C.

George W. Bush promised to do it. Nancy Pelosi promised to do it. Barack Obama promised to do it. "Changing the culture in Washington" is the bottled snake oil of electoral politics. It is the Hope and Change greasing the skids that lead us straight back to an iron-clad political equilibrium that takes our freedoms and our dollars for the servicing of their needs.

And many of us fall for it. Conservatives and Libertarians want to believe it. Independents, Republicans, and Democrats fall for it, too. Working people who don't give a damn about politics and simply want to be left to live their lives hoping that the government will do those few things it should and otherwise stay out of their lives fall for it.

It's the cynical politics of the same: same promises, same outcomes. Challengers running against seated senators do it. Even incumbents do it, by rote, fully expecting to close another sale of this magical political elixir to returning customers: "I promise not to do what I did; to not use the appropriations process as my personal reelection fund; to not kowtow to the advantaged rent-seekers who use their access to game the market to their advantage; to not feed taxpayer-financed special interests at the public trough."

But they do. They always go back to doing what they did before.

And we leave them alone to do it again. We go back to what we were doing before the last time things in Washington got so bad that it drew us out of our homes to the voting booth to push back against a political process noticeably out of control. Inevitably, the political class is again left to its natural tendencies, to hand out the concentrated benefits that will buy their next election and distribute the dispersed costs unnoticed by the voting public.

We are all like Charlie Brown, standing poised to take one more run at Lucy's perfectly placed football. We hope and we believe even though we know we should not and then we find ourselves flat on our backs looking up from the ground, cursing the politicians.

By not showing up, by not pushing back, we allow even the best public servants to be manipulated by the privileged interests inevitably drawn to plunder the riches of big government. They too will fail and fall without our support and encouragement.

Somewhere along the way, we abandoned George Hewes and the other citizen patriots who took to the streets in defense of liberty in 1773. We abandoned the traditions of citizen activism that made the American experiment a reality. We left the streets to be controlled by advocates of bigger government.

We accepted the naive notion that public officials left their self-interests at the door the moment they took the oath of office. We took solace in the belief that our ideas were so superior, so inevitable, that all we had to do was clearly explain our policy wisdom to government officials and they would eventually come around to the right way of thinking. They would do the right thing if someone just explained how to do it.

We now know the world just doesn't work this way.

And then we fell into what I call the "Benevolent Despot" trap. If we just elected the right people to public office, they would do the right thing regardless of the political consequences. We have all slipped into this way of thinking, waiting for the perfect leader to take charge and drive the right reforms of big government from the top down.

These, by the way, are the same assumptions that socialists and progressives and other advocates of big, benevolent government employ to solve social problems. Angry about the unholy collusion between banking committee chairmen and big banks? Give Congress even more power. Frustrated with the bureaucracy and seemingly arbitrary power of big insurance? Put a benevolent government in charge of a more rational rationing of your health care.

When Jann Wenner, the hopelessly liberal founder of *Rolling Stone* magazine, was interviewing Bob Dylan in 2006, he was told that politicians cannot be counted on to solve problems. To him, the statement seemed absurd.

"Who is going to solve them?" Wenner asked, incredulous.

Dylan replied, "Our own selves."

He's exactly right. If we want things to change, we need to look to our own selves to get the changing done. The culture in Washington won't change. The politicians who promise to change the culture in Washington won't change. When you think about it, that has never been the answer. George Hewes and Mary Rakovich knew it was not the answer.

We need to change the culture in the other Washington—the real Washington—in Tacoma, in Seattle and Yakima and Everett and Olympia. We need to change the culture in Jacksonville, and in Philadelphia and Evansville, Little Rock and Houston. We need to change the culture outside the D.C. Beltway, in America. We need to change things, starting in your hometown.

That's why the Tea Party movement is so different. Unlike past political uprisings against a political establishment run amok, this is a revolt from the bottom up. It is built on a coherent, unifying set of values, American values that go back to the revolutionary traditions of our founding as a nation. It is connected via the social networks of the Internet. It is built around traditions of respect and humility and hard work.

We welcome everyone from every walk of life to join our cause. We pick up our trash. We protest peacefully but insistently. This decentralized grassroots revolution has gathered disparate citizens and turned a gathering crowd into a cohesive community.

This is not a political party; it is a social gathering. Any activist will tell you about the essentially fun and celebratory nature of any Tea Party event. It's like a tailgate party before a football game or the annual family picnic. I am reminded of the sense of community you used to experience in the parking lot before a Grateful Dead concert: peaceful, connected, smiling, gathered in common purpose.

At the 2010 D.C. Tax Day Tea Party, we had some forty thousand people gather on the National Mall right under the Washington Monument. It was a typically joyous gathering despite the name-calling of our critics and in spite of the serious challenge we face as a nation. The press desperately wanted to report otherwise. They wanted to find a problem or a bad actor. So when

NBC reporter Kelly O'Donnell questioned a black participant that day she started with a not-so-subtle observation. "There aren't a lot of African American men at these events," she said. "Have you ever felt uncomfortable?" He responded, "No, these are my people, Americans."

Tea partiers have successfully taken their nascent movement from what I call "political space," a space regularly populated by a tiny percentage of the American people, to a broad cultural space, where the rest of America lives. This is the difference between a canned stump speech and a Grateful Dead concert. It is a community in the fullest sense of the word.

And that makes it sustainable. It means that we have the opportunity to change the culture in Tacoma, Washington. It means that this community will be there the day after the first Tuesday in November, holding a new crop of elected officials accountable not just for their promises but for their actions.

What an opportunity. Let's go for it.

MATT KIBBE

Appendix

FREEDOMWORKS GRASSROOTS ACTIVISM TOOLKIT

Contents

Scenes from the September 12, 2009, March on Washington.
Photos by Michael Beck

M any of you may be new to the political process.
You have probably never considered yourself to
be a community organizer. But in the last couple of
years, thousands of Americans have become true leaders in their communities, organizing protests and town
hall meetings, lobbying their state and federal legislators, and leading get-out-the-vote efforts for limited-
government candidates in elections.

No one is born an expert organizer, but we can all
learn from one another by sharing best practices and
pitfalls to avoid. In the next few pages you will learn

the basics necessary to become an effective advocate for limited government. We will describe how to organize a protest, be effective in town hall meetings, interact with the media, utilize online tools, hold events, raise money, recruit members to your group, and other essential information.

Remember that we are following the tradition of the original American community organizers, the Sons of Liberty. These grassroots Americans helped lead a campaign to build public support for the American Revolution, and were the brains behind the original Boston Tea Party in December 1773. As they understood so well, it does not take a majority to prevail, but rather an irate, tireless minority keen to set brush fires of freedom in the minds of men.

We hope that you will find this guide useful to you and your local group. There is also an online resource for those who want to learn more at www .freedomworks.org/manual.

1. THE CHAPTER CONCEPT

The chapter concept is designed to achieve the goal and mission of FreedomWorks: lower taxes, less government, and more freedom. Therefore, it is important that you know and use these methods when building your club.

The chapter concept allows activists to function in small groups so they can easily relate to one another. It promotes a sense of unity and motivates members to take action in the political system. We have state chapters, county chapters, and town chapters all across the country. You can become a part of this network of activists by creating your own chapter and networking with others in your county and state.

ACTIVIST SPOTLIGHT: BEN TESSLER

On April 15, 2009, Washington, D.C., real estate agent Ben Tessler attended his first Tea Party rally. "I was shocked to see nine or ten people I knew," he recalled. "These are normal people who feel as anxious as I do for the country. All felt helpless and at the same time, all felt a great need to be there at that rally just to be doing something."

Ben went home feeling revitalized. He forged deeper relationships with his friends he saw at the rally. He also began building a contact list and pushing the limited government and fiscal responsibility message whenever and however he could. Although it seemed like more and more people were picking up on the message, he wasn't sure exactly how to bring things to the next level.

At a July 4 gathering, Ben met Brendan Steinhauser of FreedomWorks. "Brendan and I exchanged contact information and kept in touch. When I found out about the 9/12 march on Washington, I volunteered to do anything I could to help. I had to be involved and not let someone else fight my battle. There is nothing more important than fighting for family, country, and faith. They are all tied together."

Fired up for the march, Ben went back to work on his e-mail list. He and his friends began to send out articles and video clips to those on their contact lists. As the messages were forwarded and became viral, Ben's list continued to grow.

"Our local group positioned itself as a kind of a first responder," he said. "We put ads in local papers, wrote letters to the editor, and tacked flyers on the community boards at coffee shops and pizza parlors. We put signs on major streets and hung banners on overpasses and busy intersections."

What began as an experiment for Ben had become a passion, consuming his time, and growing in directions he never imagined. "I could sense great momentum building up to the march. Citizens were waking up, feeling the same sense of urgency I felt. The mood was changing. There were more of us than anyone realized."

CHAPTER ORGANIZATION

Mission Statement

Each chapter should have a clear mission statement consisting of one or two sentences that succinctly explain the explicit goals of the group and how they will be accomplished. The mission statement provides an important roadmap for sustained effectiveness months and years into the future. When issues arise, it's often helpful to refer back to the mission statement to decide on a path to take.

A smoothly operating chapter has structure. Each chapter should have a chapter leader, communications director, legislative liaison, membership director, and social director. It is recommended that these positions be delegated to reliable individuals in each chapter. However, if the chapter is small, multiple duties can be delegated to a few individuals.

Chapter Leader

The volunteer chapter leader will be responsible for the operation and conduct of the chapter's members. Fund-raising will also be the responsibility of the chapter leader. This person is the public face of the organization and should possess excellent personal relationships, leadership, and organizational skills. Often this person is referred to as the chapter chairman or president of the chapter.

Communications Director

The press officer should be responsible for internal and external communications. Press releases, editorials, and letters to the editor are examples of external communications. E-mails and correspondences that remain in the chapter are considered internal communications. This person should maintain the chapter's Web site or blog. Also, the communications director should try to link the chapter's Web site with other local blogs and think of ways to increase traffic to the Web site or blog. This person would also manage the group's Twitter feed, Facebook outreach, Ning network community, and other online networks.

Legislative Liaison

The legislative liaison should have strong knowledge about the local, state, and national political environments and processes. It will be his or her job to keep track of legislative issues, provide updates to the group, build relationships with legislative champions, and set up activities like lobbying visits, phone call and e-mail campaigns, and other campaign activities. The legislative liaison should know what bills are being introduced, who the cosponsors are, and whether to support or oppose the bills. This chapter officer is critical when it comes to making an impact on the public policy process.

Membership Director

The membership director will maintain membership lists and ensure that all members are receiving the proper materials. Furthermore, the coordinator will work with our grassroots manager to access the FreedomWorks database for membership data. This chapter officer is in charge of recruiting and retaining members for the local chapter. It is his or her job to look for opportunities to greatly expand the membership list through creative efforts and good old-fashioned peer-to-peer recruiting. It is important for a group to continuously grow over time in order for it to remain a political force in the community.

Social Director

The social director's job is to make sure that the chapter is having fun while getting work done. This person will plan events that build a sense of community in the group: think about things like movie nights, soirees, parties, and other fun social events. It's important that the members of the chapter enjoy what they are doing, as well as feel like they are making a real difference in the public policy debates.

RUNNING A CHAPTER MEETING

Once the chapter's leadership is in place and organized, the next step is to host a chapter meeting. Chapter meetings should be held on a regular basis. They should also inform and motivate members to take action. Chapter meetings should be simple, short, and informative.

The purpose of a chapter meeting is to make sure that everyone knows about upcoming events and campaigns. But meetings are also a great place to invite special speakers or to screen a film. Meetings should be held regularly so that it becomes a part of the members' ordinary lives. If there are times when the group cannot meet due to timing, weather, or other reasons, consider hosting a conference call for everyone. Many groups have used conference calls to great effect, and you can feel free to utilize FreedomWorks' conference call system.

> *The success of a political event or meeting will hinge on the people within the organization planning the event. A dedicated core group of people is critical to the success of a meeting. These people are typically recruited through networking and from prior meetings, events, and projects. The best people are the ones who exhibit a strong desire to work to implement our ideas and policies. They will tend to be the true believers who are excited and impassioned by the opportunity to help advance our agenda.*
> —RICK REISS, TEMECULA, CALIFORNIA

Chapter meetings should typically last no longer than an hour or so. People lead busy lives and you don't want them to stop paying attention and leave early. Cover as much as you can and then leave them wanting more from the next meeting. Be sure to have someone run the meeting who is good at moving things forward. Try to save most of the debating and rhetorical questions for the very end of the meeting. You need to stay focused and get things done before you can open the floor up to banter, which is something that will happen. Manage your meeting effectively and stick to your schedule. Your members will thank you for it.

2. NETWORKING AND BUILDING COALITIONS

If your chapter is going to succeed, networking must be practiced at the local and state levels. Networking with other organizations allows your chapter to focus on what it does best, which is to mobilize citizens to take action in the political process. Other organizations will be able to assist you by providing information, resources, and people who may become members of your group.

Building coalitions is essential to leading a successful chapter. As a chapter leader you can become a member of other groups that have similar values. Coalition-building provides another means of recruiting people to your group and spreading the message in your local community. One trend that is gaining momentum is the creation of federations of limited government groups at the state level. For example, local groups have opened up the lines of communication with one another, and local leaders talk regularly on conference calls and via Google groups to stay in touch about projects everyone is working on. This model has worked with great success in many places and should be replicated around the country. No one person or group needs to be in charge. The purpose of the federation is merely to keep the lines of communication open and to allow for greater cooperation between like-minded groups.

While I have e-mail contact with most of the Tea Party leadership throughout the Bay Area, each group is wholly autonomous. In areas where there is great distance this makes sense, but in smaller geographic clusters, coalitions and larger protests should at least be one more tool in the Tea Party toolbox. Larger protests offer many advantages that scattered, smaller, local rallies cannot: precious resources can be streamlined, larger numbers in one place attract concentrated media attention, and the message of the day can be delivered with greater effectiveness. This kind of coalition should never take away from local groups acting autonomously in their own best interests, but occasional collaboration would indeed add impact. Our group has undertaken several novel steps in recent months to foster cooperation and increase exposure among Bay Area conservatives, groups, and candidates.

—SALLY ZELIKOVSKY, SAN FRANCISCO, CALIFORNIA

When building coalitions remember that all groups can keep their own identities and leadership while at the same time staying in touch with the other groups around the state. There should ideally be a meeting in every county in America where limited government activists can come together to share ideas, collaborate on projects, and network with one another. The location of these coalition meetings could change from town to town within each county, to make sure that every group has a chance to host one.

3. HOW TO ORGANIZE A TEA PARTY PROTEST

PICK A LOCATION, DATE, AND TIME IN YOUR TOWN. WE suggest Main Street at an intersection with lots of traffic.

TELL YOUR FRIENDS, FAMILY, COWORKERS, AND EVERYONE else you know about the protest. Build an RSVP e-mail list so that you can provide quick updates if something changes. You should also create a Facebook group so that the group members can communicate with one another.

MAKE FIVE TO TEN SIGNS WITH LEGIBLE SLOGANS THAT send a clear message to the public and the media. Write in big letters.

CALL YOUR LOCAL TALK-RADIO HOSTS AND ASK THEM to announce the location, date, and time on the air for a few days leading up to the protest. Send a letter to the editor of your local newspaper announcing the protest. E-mail the bloggers in your area and ask them to post a notice about the protest.

WRITE A PRESS RELEASE AND E-MAIL, MAIL, AND FAX copies to the local TV stations, radio stations, bloggers, and newspapers. Call the reporters who cover local events or politics and leave messages on their voice mail.

ON THE DAY OF YOUR PROTEST, SHOW UP WITH YOUR group; be loud, visible, and happy; and engage the public. Wave your signs, make lots of noise, and move around to get attention. If reporters interview you, give them some good sound bites for their stories. Stay on message and keep your answers short and coherent.

BRING SIGN-IN SHEETS TO CAPTURE THE NAMES, E-MAIL addresses, and phone numbers of those who attend the protest and/or say that they support what you are doing. You will then have a big list of people who can plan the next, much bigger and louder event. Also, bring handouts with one page of quick facts about why you are protesting in the first place.

ADD YOUR PHOTOS, VIDEOS, AND AN AFTER-ACTION report to your Facebook group, and send this material to the bloggers and reporters whom you originally contacted. Ask them to post the photos, story, and video.

THANK EVERYONE WHO ATTENDED VIA E-MAIL AND PHONE, and set up a meeting to plan your next event. Now you have a list of people in your community who can help make the next protest huge. Encourage everyone to commit to bring at least one friend to the next protest.

ORGANIZE A CARPOOL AND GO FIND A FRIEND IN YOUR neighboring town or county and help them organize a protest there. You and your people are now veterans and should be able to keep the momentum going around your area.

ACTIVIST SPOTLIGHT: BILLIE TUCKER

"Where are our leaders? Where are those who are responsible?" asks Billie Tucker, an entrepreneur from Jacksonville, Florida. It's a question she has pondered the past two years.

"I came to the realization that no one in Washington, not the president, not the treasury secretary, not the candidates for president, and certainly no one in Congress had any clue what was going on or what to do. I trust these guys to mind the store and when the store is on fire, they didn't even know where to find a garden hose. As a parent, it reminded me of how children behave when they don't understand something: They make it up."

Billie happened to have her TV on CNBC during Rick Santelli's rant.

"He said what I was thinking!" she recalls.

Later that day, a friend called and asked Billie if should would be interested in helping plan a Tea Party event. Less than twenty-four hours later, more than seventy-five activists gathered for the first Jacksonville Tea Party.

What began as a small list of phone numbers later became the First Coast Tea Party, one of the largest and most active groups in the country. At the tax day protest in Jacksonville on April 15, the unlikely leader had a close look at what she helped create. "Crowd estimates told us that 5,000 people showed up that day," she said. "It was hard to take in, especially as we saw ourselves on news reports across the country."

Over the summer the First Coast Tea Party also chose to focus on congressional town halls. Intimidated by the prospect of active, well-informed attendees at live events, several Jacksonville area representatives opted for virtual meetings that connected voters by phone.

"We decided we would hold live town halls whether a congressman was there or not," Billie said. "We taped their pictures to empty chairs on-stage. It did a good job of reflecting how we felt about communicating to Congress—talking to an empty chair."

4. TRADITIONAL MEDIA

There are two types of media coverage: earned and paid media. Earned media is free exposure for your organization. It is when the media shows up at an event and covers it or publishes your opinion. The content of earned media is up to the reporter and is difficult to control. Earned media is more credible and—best of all—it's free.

Paid media, on the other hand, costs money. Paid media will say what you want and be designed according to your budget. Paid media includes newspaper ads, radio ads, TV ads, online ads, and other types of marketing efforts.

Our simple rule of thumb at FreedomWorks is that you should try to get earned media first and only use paid media as a last resort. If you do your job and find creative ways to get attention for your group, you shouldn't have to use very much paid media, thus saving yourself a lot of money. Following are some tricks of the trade when trying to get mentioned in the media.

LETTERS TO THE EDITOR

Letters to the editor provide citizens with the opportunity to comment on articles and editorials appearing in their local newspapers. Studies show that people read the letters section of newspapers more than they read the editorials by journalists.

Moreover, letters to the editor are widely read by community leaders and lawmakers to gauge public sentiment about current events. Here are some helpful guidelines to follow when crafting your letter to the editor:

Write legibly. Include your name, address, and telephone number. Papers often call to verify authorship. Newspapers generally will not print anonymous letters.

Address your letters to the "Letters Editor" or "Dear Editor."

Be brief and specific. Letters should never exceed one page. State the purpose of your letter in the opening paragraph and stick to that topic. If your letter pertains to a specific article or editorial, identify it accordingly. Try to keep your letter under two hundred words. Always adhere to the paper's guidelines, which should be clearly stated on the editorial page.

Write nothing but the truth. Do not include false information or dubious sources in your letter. Mentioning studies and statistics in your letter will enhance its effect, but don't overdo it. Your message can become lost in a sea of figures.

Write about current issues. Stick to debates going on right now. Respond promptly to tax-increase stories and editorials. Write in support of pending tax cuts or against pending tax increases.

Look at published letters in the paper you are submitting to; they will usually have a format you should follow.

Don't become discouraged if your letter is not published. Most publications receive more letters than they can print and will often print one letter as a representative of others. Most important, keep trying. Unpublished letters are still read by the editors and can help them determine which topics should receive more attention.

OP-EDS

Editorials appear in most newspapers and are vehicles by which citizens can make extensive comments on articles and policies. Like letters to the editor, editorials are placed on the editorial page. They are commonly referred to as op-eds because they sometimes appear on the opposite page from the newspaper's editorial. They often have the ability to reach a large audience. When writing an op-ed, be sure to keep your piece concise and include specific information. The average length of an op-ed should be between four hundred and nine hundred words. Check with your local paper concerning length requirements.

Your op-ed should be timely, concise, and to the point. Make sure that you drive home one or two main messages in the piece. Organize it well and make the strongest case for your point of view. Utilize every word to the fullest and don't waste time with ad hominem attacks or other distractions. Get right to the point and flesh out your arguments.

When arranging the publication of an op-ed piece, a phone call to the editor can be helpful. Leaders of organized groups often have a much better chance of getting an editorial published in the local newspaper than the average person. Some local newspapers have cut back dramatically on the acceptance and publication of guest op-eds. But there are plenty of opportunities to submit editorials online. Generally, you will have a much better chance of publishing a letter to the editor than an op-ed, but this shouldn't discourage you from submitting one when it can help you in the battle of ideas.

RADIO AND TELEVISION TALK SHOWS

Calling the talk shows in your area is a great way to help get your message across to thousands of listeners for free. Call your local radio and television stations and ask if they have open forums—talk shows where callers can discuss any subject with the host. If so, try to get on the air to make short, concise, positive statements about limited government. If there is currently a bill making its way through the legislative process, the host may keep the topic on the air for several minutes. If not, then at least you can take comfort in knowing that your brief statement in support of our position was heard by the station's listeners.

You can also call talk shows and ask the producer if there are any scheduled shows coming up that will discuss our issues. If one is scheduled, try to get a representative booked to appear on the show. Be sure to monitor your local radio and television stations and participate in these shows and alert fellow volunteers so they may do so as well.

Most communities have a local radio station. In this case the smaller, the better. Small stations do community announcements as a public service, so send them the information. If they have a local talk show, reach out to the host. Do not attach signs to poles or place them in public rights of way. You want to maintain a positive image. When people begin to contact you (and they will), enlist their help and get their e-mail address. E-mail them a copy of the flyer and

*ask them to call their friends. You can use window
chalk (liquid) on your car windows. I have a trailer I
use every day and a big sign that goes on the back
about a month prior to an event.*
— ROBIN STUBLEN, PUNTA GORDA, FLORIDA

FreedomWorks often has opportunities to book activists who we work with on national media outlets like Fox News, MSNBC, CNN, PBS, and the broadcast networks. If you prove to be an effective media voice for limited government, we will be happy to include you in future opportunities. Producers are always looking for good spokespeople who can go on the air live. This is a great opportunity for those of you who want a chance to do something like this. A number of FreedomWorks activists have appeared on national radio and television shows, and some have even been profiled by outlets like CNN and Fox News.

BUILDING MEDIA CONTACTS

You will find contact information for television, radio stations, and newspapers across the country on a national and local level online. It is highly recommended that you build a media list of your own for your group. Keep this information handy, as you'll want to refer to it when you're ready to mail your letters, place your telephone calls, or send e-mail.

Every local leader should have a big list of the reporters and producers who have covered local, state, and national politics. Many reporters are assigned to covering political events in the city or state in which you live. Get to know them personally and communicate with them often. They need you as a source just as much as you need them to write a fair piece about your group and your events. As frustrating as the mainstream media can be, there are good reporters and producers who are honest, objective, and looking for the true story. Work with these folks and you can help shape the narrative about our movement. I'd suggest offering to take your local beat reporters out to lunch or for coffee to get to know them. You'd be surprised at how few people actually do this, which is a major missed opportunity. Good journalists will take you up on your offer because they need to know their subjects very well in order to cover them correctly.

Try to focus on building not only the list of media contacts but also the quality of the list. This will go a long way toward making you a more effective communicator through newspapers, radio, and television. We can and should get our message out on traditional media outlets. To throw up our hands because of media bias would be foolish and counterproductive.

ACTIVIST SPOTLIGHT: BOB MACGUFFIE

Bob McGuffie was on a ski trip in New England when he heard Rick Santelli's rant. By then, he had been organizing protests and connecting activists for some time and was prepared for the tsunami of activism unleashed by the instantly famous outburst. A native of Queens, New York, Bob is blunt and unapologetic about his views.

He is a cofounder of the political action group Right Principles (www.rightprinciples.com), a clearinghouse for economic and constitutional research and a gathering place for activists. As hits and e-mails to the Web site grew during the TARP fiasco, Bob realized he was at the center of something special. Technology was empowering individuals across the country to connect with one another, share ideas, and plan events.

After the April 2009 tea parties, Bob and his fellow Connecticut activists decided to focus on town hall meetings with members of Congress. "I thought we ought to go down and challenge them," Bob said. "Representative Jim Himes had a town hall coming up—on my wedding anniversary of all days—but I wanted to start pushing back on all the happy talk. I ended up challenging him on the national debt and on the cost of the stimulus."

Excited about the potential for similar results around the country, Bob penned a memo called "Rocking the Town Halls." The message was straightforward: get your network together and come armed with facts; spread out across the hall and get to the microphone; challenge the political talking points. The memo was read by thousands of activists who used it to organize participation in town halls and challenge lawmakers.

Bob's growing network of supporters combined first-time activists with veterans from a wide variety of existing organizations. A July barbecue cookout boasted representatives from no fewer than twenty-three distinct groups including Libertarians, Glenn Beck–inspired 9/12ers, second amendment defenders, and more. "We are a loose coalition united around the principles of small government," Bob says of the political kaleidoscope in Connecticut. "We knew we would be strong if we worked together. We're all travelers in the same movement, so we informally established the Connecticut Grassroots Alliance."

Bob points out that the network represents Connecticut. "We have both urban and rural, rich and poor. We're your neighbors and represent all the backgrounds that make up Connecticut. There is a myth that small government conservatism is only popular in the American south. Well,

I've got a network here in New England that dispels that myth."

5. SOCIAL MEDIA

There has been much written about Obama's online campaign leading up to the 2008 presidential election. The greatest strength of his campaign was that he was able to meet people where they were through social media. People may or may not log on to Web sites daily, or even weekly—but they're sure to check their Facebook and Twitter accounts on a daily basis.

The online world can be immensely intimidating to someone who is just figuring out social media. There is a constant influx of information and content, and organizing and making sense of it all seems overwhelming. The trick is harnessing all of that information and using it to engage other activists. It has helped the Tea Party movement grow and connect in new ways, creating a large, powerful, and sustainable movement. The power of these tools is undeniable, and using them may not be quite as difficult as you think.

NING

Ning is a site that allows users to create their own social networks. Leading up to the 9/12 Taxpayer

March on Washington, D.C., FreedomWorks created our own network to help people connect and coordinate. Local groups were started, people were able to connect and arrange transportation and share stories. A Ning network is very effective for organizations with members across the country. Smart Girl Politics began as a Ning network, and the site still serves as a hub, with members blogging, posting photos, joining their state groups, and sharing events.

If you feel that a Ning network would benefit your group, you can go to ning.com and create your network. Or join FreedomWorks' Ning community at teaparty.freedomworks.org. This fast-growing network has quickly become an organizational hub where you can join your state group, plan and post events, network, and discuss important issues with others near you.

FACEBOOK

Facebook is the most widely used social media platform. It's incredibly user-friendly and easy to learn. Here's a step-by-step walk-through to getting started. Be sure to join us on Facebook at www.facebook.com/freedomworks.

STEP 1: Signing Up and Creating Your Profile

Getting started on Facebook is free and takes about five minutes. You just go to Facebook.com and enter your name, e-mail address, sex, and birth date and click SIGN UP! Use your real name. It's tempting to use the name of an organization or group—but that is what the Pages are for. Your profile should be about you.

A big concern for many people new to Facebook is privacy. This is a valid concern—after all, it is the Internet. Nothing is "private." Facebook, however, has privacy settings that you should take full advantage of. The safest way to approach online privacy, however, is to simply not post things you're not comfortable sharing. You have the ability to hide your age, location, and networks. Don't want people to know where you work? Don't post it.

STEP 2: Connecting with Others

When you sign up, Facebook will prompt you to search for people by e-mail. Here you can enter in your e-mail login information, through Gmail, Yahoo, or whatever e-mail service you use. It will search by e-mail for people you know who are already on Facebook, and you can automatically add them as friends. Of course the search isn't perfect. It will miss people. You can search by name and e-mail address to find your friends, classmates, neighbors, coworkers, and family members.

There are several ways you can approach your Facebook account. Most people will only add and accept requests from people they know. This is the safest route, because really, who wants strangers to have access to pictures of their kids? However, as an activist, there can be a much broader use for Facebook. The other approach would be to accept other people who want to network with you—maybe a grassroots leader from a nearby city or even across the country. Share information and help promote one another's events to your own networks. The more, the merrier!

STEP 3: Using Fan Pages

This is really useful if you have an organization or group you'd like to promote. There are several benefits to Facebook fan pages, not the least of which is that there is no cap on the number of people who can "like" your page. A basic profile has a limit of five thousand "friends" . . . which is plenty if you're using Facebook to communicate with family and friends. However, you may want your reach to be much broader.

The easiest way to grow your group's page is by inviting your friends and having your friends invite their friends. Have a bit of a budget? Buy a Facebook ad!

Facebook ads are relatively inexpensive and can also be incredibly effective since they can geo-target your audience. You can choose to have your ads show up only to people in your city, state, or region.

TWITTER

Twitter is slightly more intimidating for many. Some just plain don't get it. There is an incredible amount of information out there, and a lot of it may very well be a waste of your time. However, it is also important to embrace the power of the medium and make it work for you.

Twitter served as an incredible catalyst in planning and organizing the beginnings of the Tea Party movement. It connected leaders from across the country and launched several grassroots organizations. Our Twitter handle is @freedomworks.

STEP 1: Getting Started
All you need to provide is your name and e-mail address. It's a five-minute (if that) sign-up process. Suggestion: Use a real photo for your avatar, unless you are an organization. If you use a logo or cartoon, generally people assume you're (1) spam, unsolicited communication, or (2) a troll, one who signs up only to insult and pick

fights. If you're concerned about identity, you can use something that will make people laugh.

Your bio is extremely important on Twitter. It's the only thing that people will have to figure out who you are, and you only have 160 characters to do it. Start with something basic. List which organization you're with (if there is one) or what you're passionate about. At FreedomWorks, we simply list our tagline—Lower Taxes, Less Government, More Freedom. Just six words, but it gives people a clear picture of what we're working for and what our mission is.

STEP 2: Who Do I Follow? And How Do I Get People to Listen?

The best way to figure out who to follow is to go to people you admire and follow the people who they're following. Pick an active Twitter personality you are familiar with and pay attention to who they are following. Watch who they are replying to and follow them. In addition to following individuals, pay attention to the lists they create. Many have created lists of people they follow, which are all linked on the right-hand side of their profile page.

How do you reply to someone? Simply put an @ before their name. For example, if you would like to bring something to the attention of the FreedomWorks account handler, you would start your tweet with @ FreedomWorks, and the tweet would be directed to us.

Retweeting is also important. When someone says something interesting and you want to share it, simply use the retweet feature and it will be sent back out to those following you, preceded by an RT and the name of the person who composed the tweet.

The best way to get people to notice you, however, is to engage them in conversation. If they follow you and are interesting, follow them back. This opens the line of communication. Twitter is about dialogue—it is a conversation starter. To use it to its potential, you need to listen as much (or more) than you speak.

STEP 3: Third-Party Applications

One of the best features of Twitter is that it's completely integrated into your phone and computer. The web version is always available, but by downloading desktop applications, you can create groups, manage multiple accounts, and organize your Twitter feed.

Here are some of the most popular desktop applications:

TweetDeck
Splitweet
Hootsuite

Seesmic

Twhirl

Twitter can also be managed via phone. BlackBerry and iPhone both have great Twitter apps.

BLACKBERRY

Ubertwitter

Twitterberry

TinyTwitter

IPHONE

TweetDeck

Hootsuite

Echofon

Don't have a smartphone? You can link your phone to your Twitter account and update via text.

To get the full effect of Twitter, keep it at your fingertips. It's incredibly effective to cover rallies or any other event. Send photos from the middle of a crowd and keep people informed!

The biggest asset of Twitter is communication and information sharing. This can go viral with a click.

To be effective, be consistent and be engaging. People want new information.

THE BLOGOSPHERE

STEP 1: Starting Your Blog

Blogs are a huge component of the online news world. Today's bloggers are investigative reporters, digging up dirt, and pointing out political trends. Bloggers are shaping the conversation and are a critical part of the political narrative.

The lines have been blurred between blogger and journalist—bloggers *are* the new media.

Starting a blog is simple and free. Platforms like Wordpress.com, MoveableType, and Blogger.com make it a point-and-click process. Just sign up, choose a catchy name and a template for your site, and you're free to start writing. Pick something you're interested in to write about. Maybe there's a particular candidate you want to support or a policy you're interested in covering. Find the information and start posting it.

STEP 2: Getting Your Blog Noticed

Linking is maybe the most important way to get your blog noticed. Has another blogger covered a story that you can add something to? Post a link to their piece, quote the relevant part, and then add your response. Some sites will display the links to people who have linked their story—they're called trackbacks and are also a way to drive traffic to your site.

SEO (search engine optimization) is important. Choose headlines that are descriptive, with words that people will use in their Google searches. For Word-Press, there is also a plug-in that allows you to create a separate headline and description for the Google searches, creating more search engine traffic to your blog.

STEP 3: I Don't Have Time to Blog. Why Should I Care?
Blogging is time-consuming, and it isn't for every-one. There are, however, ways you can contribute to the conversation without maintaining your own site. At RedState, everyone can be a diarist. Have a story or some coverage of an event? Post it as a diary. Some-times the editors will promote it to the front page.

Also, if you don't know how to get attention for an event or story, pass it on to a local blogger. Chances are, people who matter will read that person's site. An ex-ample would be a blogger we know from Kansas. Now, he doesn't have the traffic of one of the major sites, but he has a feature that allows people to subscribe to his blog by e-mail. His subscribers include the mayor, every member of the city council, and the staffs of both. What this means is that when he voices concern or support for particular policies or initiatives, the people who can affect change are listening.

YOUTUBE

YouTube is an incredibly useful tool. Video has become the most powerful medium available in the online world. YouTube is a simple way to share any footage and television clips, or your own montages, slideshows, and so on, with the world.

You can start your own YouTube account for free. It's linked to your Google account, so the people in your contacts will show up as suggested subscriptions— subscribe! It creates a feed and you can keep up with new videos posted by your friends.

Upload original videos for your subscribers. Make sure to take them in the info area while you're uploading them so people will find them when they search for those terms. The title and description for the video are important—make them as clear as possible so more people will find your video when they search for a particular topic.

Even without making your own YouTube account, the platform is a great resource. You can link videos, share them on Facebook, Twitter, and your blog, and promote other videos if you think they are interesting and useful.

Many people use YouTube as an information resource, finding their news and doing research by wading through the videos. It's second only to Google as a search engine. Having news clips, videos of candidate speeches and interviews, footage of events, and other media clips available can prove invaluable.

ACTIVIST SPOTLIGHT: C. L. BRYANT

Activism has always been a part of C. L. Bryant's life. He has long been a champion for civil rights, marching throughout the 1960s and 70s for freedoms many take for granted today. He has marched with Republicans, he has marched with Democrats, and on occasion he has marched alone. He is proud of what he has achieved. But as the years passed, he became disillusioned as many he worked with became more concerned with left-wing partisan politics than the movement.

"I found myself increasingly looking to conservatives," he said. "Racism is a collectivist concept and I believe we should all be judged individually by our character."

In his heart, C.L. is a proud American who believes our best days are still ahead of us if we stay true to our founding principles and committed to the values that made us strong. "My early activities were inspired by Frederick Douglass," he explains. "Douglass believed in equal opportunity, individual freedom, and liberty. We should strive to live up to his courageous example and fight for what we believe is right."

For C.L., the Santelli rant was a call to arms. He started hearing rumors about upcoming local tea parties and found an event coming up in Shreveport, so he decided to go and see for himself. The Tea Party movement had found a new member.

"The organizers thanked me and asked me if I'd like to say a few words," he explained. "After the confrontation, I was fired up." C.L. spoke again in Shreveport at a July 4 event and again his words were inspiring to the Tea Party nation. Someone in the crowd videotaped the message and it was soon forwarded to activists across the country.

"I started to get questions from the press," C.L. said. "They wanted to know why I would be involved with the movement." The premise of the question insulted C.L. It was based on the assumption that an African American couldn't possibly identify with a conservative, small government message.

Undaunted, C.L. started to attend more events. Brendan Steinhauser of FreedomWorks saw footage of his speeches and invited him to appear on the main stage at the 9/12 march on Washington.

"I said that Americans, regardless of race, gender, or circumstance, would not be ruled over. Our freedoms come from our Creator, not the government. I believe politics has become about walls of divisiveness to keep us separated. I wanted to invoke the spirit of Reagan to tear down these walls."

6. LOBBYING ELECTED OFFICIALS

Everyone is familiar with the slick, well-funded special interest lobbyists who constantly roam the halls of congressional office buildings in Washington, D.C., and your state capital. But in the eyes of elected officials, the most effective lobbyists are their constituents. Politicians are keenly aware of the fact that it is their constituents who hold the keys to their political futures. As a result, constituent concerns are of prime importance to politicians. The best way you can affect the outcome of legislation is to directly communicate your views to your lawmakers.

We believe that government goes to those who show up. If you aren't the ones showing up in your lawmaker's office, on the phone, and in the mailbox, those that are showing up are the entrenched special interest groups that don't necessarily have *your* best interests in mind. The simple act of making a visit to your local district office, or making a phone call can make a big difference. Most offices have a formula when it comes to constituent contact. For instance, one phone call equals another hundred people who feel the same way, so your voice is magnified many times over.

Effective grassroots lobbying can be done at the local, state, and national levels. The same general principles apply, and the following advice can be read in that context.

Remember, lawmakers work for you (and since the number-one concern of all politicians is to get re-elected, be sure you communicate your concerns with your own elected officials first). To assist you, this section provides you with tips on effective communication strategies with your elected officials.

KNOWING YOUR LAWMAKER'S OFFICE

Whether local, state, or federal, the personnel of the offices of your elected officials are similar. Most legislators have a staff to assist him or her during his or her term in office. To be effective in communicating with these offices, it is useful to know the titles and principal functions of the staff. Commonly used titles include the following:

Administrative Assistant or Chief of Staff

This staff person reports directly to the member of Congress or state legislator. He or she usually has over-all responsibility for evaluating the political outcome of various legislative proposals and constituent requests. This person is in charge of overall office operations, including the assignment of work and the supervision of key staff. You should always attempt to speak to this person if you can't speak directly to the legislator. The next time you are in Washington, D.C., be sure to stop by your congressional and senate offices and pick up business cards with contact information for key staff members, usually available at the reception desk.

Legislative Director or Legislative Assistant

The legislative director is usually the staff person who monitors the legislative schedule and makes recommendations regarding the pros and cons of particular issues. In some congressional offices there are several legislative assistants and responsibilities are assigned to staff with particular expertise in specific areas. For example, depending on the responsibilities and interests of the legislator, an office may include a different legislative assistant for health issues, environmental matters, taxes, and so on. If you can't get ahold of the administrative assistant or chief of staff, the legislative director or legislative assistants are your second most important points of contact with the legislator's office.

Press Secretary or Communications Director

The press secretary's responsibility is to build and maintain open and effective lines of communication between the legislator, his or her constituency, and the general public. The press secretary is expected to know the benefits, demands, and special requirements of both print and electronic media, and how to most effectively promote the legislator's views or position on specific issues. This person is often the most sensitive to bad PR or good PR, so keep that in mind when you are trying to get a public statement of some kind from the lawmaker. Often a well-placed call to the press secretary will go a long way to making sure that the legislator faces bad PR if he or she goes against the wishes of his or her constituents.

Personal Secretary or Scheduler

The scheduler is usually responsible for allocating a legislator's time among the many demands that arise from congressional responsibilities, staff requirements, and constituent requests. The scheduler may also be responsible for making necessary travel arrangements, arranging speaking dates and visits to the district, and so on. If you are looking for dates and times of town hall meetings or if you would like to invite your legislator to speak at or attend one of your events, call his or her office and ask to speak to the scheduler. There is often a district or state scheduler as well for congresspeople and senators.

Caseworker

The caseworker is the staff member usually assigned to help with constituent requests by preparing replies for the legislator's signature. The caseworker's responsibilities may also include helping resolve problems that constituents present in relation to federal agencies, such as Social Security and Medicare issues, veteran's benefits, passports, and so on. There are often several caseworkers in a congressional office.

Other Staff Titles

Other titles used in a congressional office may include executive assistant, legislative correspondent, executive secretary, office manager, and receptionist. The legislative correspondents, or LCs, usually are the ones who write responses to constituent letters and e-mails. If you can't reach the chief of staff, legislative director, or legislative assistants, try to talk to an LC who works on the issue that you are concerned about.

GRASSROOTS LOBBYING TIPS

The most effective way to articulate your views to your elected officials and to affect the outcome of legislation is to sit down and speak with your legislators face-to-face (or with their key staff if they are not available). Usually, either one-on-one meetings or small groups is best. While these personal visits are extremely productive, they also require the most amount of planning. Here are some things to remember:

IF YOU ALREADY HAVE APPOINTMENTS SCHEDULED WITH your lawmakers, be on time.

EXPLAIN HOW THE PROPOSED LEGISLATION WILL DIRECTLY affect you. Use specific examples.

ALWAYS BE POLITE. YOU WILL NEVER CONVINCE YOUR LAW-maker or their staff with rudeness, vulgarity, or threats. Even if you disagree with the position of your legislator, be courteous and calm. There will be other issues in the future and you'll want to be able to meet with the legislator again.

FOLLOW UP YOUR VISIT WITH A LETTER. REGARDLESS OF how your meeting goes, send a letter to your legislator or the staff person you met thanking him or her for their time and reiterating the points you discussed. This gesture will help the cause and pave the way for future meetings.

In my role as legislative liaison for North Carolina FreedomWorks, I monitor the North Carolina General Assembly. When issues of interest to our more than forty thousand activists arise, I e-mail a copy of the proposed legislation and a FreedomWorks analysis. Activists visit legislators in their home districts, attend committee meetings to speak for or against the proposal (often filling every chair in a committee room), go door to door at the legislature speaking with legislators, send letters and e-mails, and make telephone calls. North Carolina FreedomWorks activists are known for virtually shutting down the e-mail and voice mail systems at the North Carolina General Assembly. But it takes coordinated efforts to make an impact in this way. That's why I think that it's important that other groups have legislative liaisons as well.

—KATHY HARTKOPF, HILLSBOROUGH,
NORTH CAROLINA

WRITING YOUR ELECTED OFFICIALS

If you haven't communicated with an elected official before, and you want to get started, the simplest thing you can do is utilize the results-oriented method of letter writing. A letter is an easy way for you to let lawmakers know your views as a voting constituent on specific issues, encourage them to vote your way, and let them know you'll watch how they vote on a particular issue and keep that vote in mind when it comes time for their reelection.

Try to keep your letter short and to the point, with just enough facts and figures to further enhance your statement. Never lie or make a statement you can't back up with evidence. Always let your lawmakers know how a specific issue will affect you; make sure they understand that you live and vote in their district or state, and therefore, what affects you may affect other constituents as well.

Always use the letterhead of your local group and identify yourself as the chapter leader when writing your elected officials. Handwritten letters are best in the age of e-mail and faxes. A handwritten letter will stick out, and usually legislative offices require their staff to send a response to all handwritten letters. One tactic that FreedomWorks has used to great effect is delivering handwritten letters to congressional district offices. Your group can collect letters from friends, families, and colleagues and schedule a time to hand-deliver them to your congressman's district office. This is a great activity for your organization to do in order to communicate your views to elected officials.

Here are the four important things to remember when writing your letters:

HOW TO ADDRESS YOUR REPRESENTATIVE: Address your letters to "The Honorable ———," and begin the letter "Dear Senator" or "Dear Representative." If writing to a committee chairman or Speaker of the House, address them as "Mr. Chairman" / "Madam Chairwoman" or "Mr. Speaker" / "Madam Speaker."

BE BRIEF, SPECIFIC, AND COURTEOUS: Ideally, letters shouldn't exceed one page, and the purpose of your letter should be stated clearly in the first paragraph. If your letter pertains to specific legislation, identify it accordingly. To make sure your letter is as productive as possible, always be courteous, even if you disagree with the lawmaker's position.

ASK THEM TO RESPOND: Always ask for a response of some kind to your letter. You'll want a hardcopy of your legislator's positions on these issues for future reference and to document his or her positions.

WHEN IN DOUBT, ASK FREEDOMWORKS: Remember that the resources of the FreedomWorks national office are at your disposal. The grassroots team always stands ready to assist you. If you need ideas on what to write, or even if you have problems locating the mailing address, contact FreedomWorks and we will be happy to assist you.

FAXING

Nearly all state legislators and U.S. senators and representatives have public fax numbers, but we are always available to assist you in acquiring a fax number that you cannot find. Faxing allows you to send a full, letter-length message to your lawmaker in a matter of minutes. When preparing a fax message to a lawmaker, follow the same basic guidelines used when mailing a letter via regular mail. You also want to make sure your fax number is clearly visible, in case your legislator wishes to respond to you via fax.

Be sure to call the office and follow up on your faxed letter. Make sure that the office received it and ask for a written response from your lawmaker. Group leaders should coordinate campaigns to fax dozens of letters to lawmakers right before a critical vote. This can be a very effective tactic in the days leading up to a vote on an important issue that you care about. These minicampaigns should also be employed when coverage of the issue picks up in the media or when there are committee meetings about the bill that you are concerned with.

E-MAIL

E-mail allows you to communicate with your legislators in the quickest and easiest manner. Unfortunately, because of the incredible volume of e-mail legislators on the state and federal levels receive, the impact of an individual e-mail as opposed to an individual letter is limited. This does not mean that e-mailing your elected officials is useless as a lobbying tactic. It just takes large amounts of e-mail on an issue to catch the legislator's attention.

Large e-mail campaigns that generate short, simple messages at a critical time (such as right before a vote or election) can have a huge impact and make great activities for your chapter. Alternatively, try to gain access to the e-mail addresses of the legislative staff. Your e-mail will more likely be read and recorded if you are in direct contact with them. Again, follow the basic guidelines for a written letter when you send an e-mail message.

ACTIVIST SPOTLIGHT:
DEBBIE DOOLEY AND JENNY BETH MARTIN

Back in the 1980s, Debbie Dooley was proud to vote for Reagan. It's been a long time since the Atlanta native has felt the same pride in the voter's booth. Over the years, Debbie had became equally disgusted with both parties. But two things changed to draw her back into the process: the birth of her grandson and the bailouts following the housing crisis.

The only problem was that she had no idea how to hold an event and did not know if she could find anyone else to join her. "I contacted FreedomWorks and found out about a conference call for folks looking to host tea parties across the nation. I got a chance to meet folks on that call who felt the same way I did—even a few in Atlanta."

One of the other participants was Jenny Beth Martin from Atlanta. The two connected and got to work.

Jenny Beth had always been politically active and was a veteran of the fight against big government. But the recession was taking a toll on the family and affected her ability to take part. Still struggling to get by, Jenny Beth eventually got involved with Smart Girl Politics, a national network of conservative women. Soon after, she encountered Debbie Dooley on the FreedomWorks call. The next day, Debbie secured a demonstration permit. In less than a week, a Tea Party was scheduled in Atlanta for February 27, 2009.

On the day of the event, rain poured on the statehouse steps. They had no stage, no sound system. They used a nearby statue as a podium. Despite all the challenges, more than 300 citizens arrived to proclaim their opposition to the reckless government spending.

Debbie and Jenny Beth threw themselves into the cause and became more involved with FreedomWorks and the growing Tea Party patriots group. In addition to a full-time job, Debbie estimated she was spending thirty hours a week on the cause. "I didn't even have time for TV anymore," she said with a laugh. "Aside from watching University of Alabama football games, of course. We have to have some priorities in life."

PHONE CALLS

As bills move through the legislative process, you will find there simply isn't enough time to write your legislators immediately before a key vote on a certain issue. When you need to get in touch with your legislator immediately to let him or her know of your support for lower taxes, less government, and more freedom, your telephone calls become the most effective means of communicating your views. During the months-long debate over Obamacare, FreedomWorks members placed hundreds of thousands of phone calls to lawmakers in Washington and around the country. The phone lines were so busy during some days that the Capitol switchboard was giving everyone busy signals. When that happens, we encourage our members to start calling local, district offices.

Here are four things to remember when calling your elected officials:

IDENTIFY YOURSELF AS A CONSTITUENT: As someone who lives and votes in the district or state, your phone calls carry the most weight. Calls to representatives outside your district or state are helpful as well, but be sure to contact your legislators first. Encourage your friends and fellow activists to call after you have placed yours.

STATE YOUR POINT QUICKLY AND CLEARLY: Be sure to limit your telephone call to one subject, and be brief and specific. Your phone call should last only a few minutes. Let your legislator know why you're calling, giving a specific bill number if possible. As with any communication with your elected officials, remember to always be courteous.

REQUEST THAT YOUR LEGISLATOR FOLLOW UP WITH A LETTER: Be sure to give your name and home address and request that your legislator follow up with a letter. You took the time to call, so ask your legislator to take the time to respond. Be sure to get the name of the person you talked to. Remember that the chief of staff, legislative director, and legislative assistants are the people who you should ask for when you call.

IDENTIFY YOURSELF AS THE LEADER OF YOUR LOCAL GROUP: Over time this will begin to resonate with the offices of elected officials as you build relationships with them. As your group grows in strength and gains credibility, the fact that you are calling on behalf of your group will mean much more. Organized activists make a much bigger impact working together than they do working alone. And your legislators know this, which is why you should remember to mention that you are involved with a local group.

PERSONAL MEETINGS

By far the most effective way to articulate your views to your elected officials and to affect the outcome of legislation is to sit down and speak with your lawmakers (or if they are not available, key staff) either one-on-one or in a small group accompanied by the other key leaders within your group. While these personal visits are extremely productive, they also require the most amount of planning to ensure success.

During the Obamacare debate, FreedomWorks activists made district office visits a key part of our strategy. Never before had so many people shown up in these local offices around the country. The staff was forced to pay attention to the groups and put the Washington offices on alert that constituents were outraged about the various health care takeover bills being debated in Congress.

Here are six things to remember when planning a personal visit with the office of your elected officials:

SCHEDULE AN APPOINTMENT: Elected officials have extremely busy schedules. To ensure that you will have time allotted for you to speak directly with your legislator, call in advance to set up an appointment. If you call enough in advance, speaking to your elected official directly should not be a problem. However, if he or she is not available due to a scheduling conflict or a last-minute problem, it is still worth your time to meet with the staff person who handles the issue that you want to discuss. Be persistent and follow up with all requests so that you can lock in a specific time, date, and place for a meeting. Be on time.

PREPARE QUESTIONS AHEAD OF TIME: Have specific questions in mind dealing with your legislator's point of view or stance on an issue. Make sure you get an answer. If you asked your question clearly and directly, you should receive a clear and direct answer. If your legislator sidesteps the issue or does not answer your question, calmly repeat it.

EXPLAIN HOW THE PROPOSED LEGISLATION WILL DIRECTLY AFFECT YOU: Use specific examples to show your lawmaker how issues affect you and the freedom of our country. If the proposed measure cuts taxes, limits government, or otherwise benefits the consumer, specifically cite examples to support this position. If you are a business owner, mention the effects the bill will have on your business and your workers. If you are a teacher, cite your experience in education and explain how the proposed bill would affect what you do. Personal anecdotes are often the most remembered and most powerful forms of communication. Sometimes the legislator will even quote you on the House or Senate floor when giving a speech about the issue.

AGAIN, ALWAYS BE POLITE: Nothing is more detrimental to a visit with a lawmaker or his or her staff than rudeness, vulgarity, or threats. Even if you disagree with the position of your legislator, be courteous, keep calm, and do not become overagitated. Also, be sure to dress professionally to convey the seriousness of your visit.

LEAVE A LEGISLATOR "LEAVE-BEHIND" AND YOUR CONTACT INFORMATION WITH YOUR LAWMAKER OR STAFF: A legislator "leave-behind" can be a short summary of the issue with key points, or as simple as a letter or petition. This will ensure that your lawmakers remember the issues you discussed.

FOLLOW UP YOUR VISIT WITH A LETTER: Regardless of how your meeting goes, send a letter to your legislator or the staff person you met with thanking him or her for their time and reiterating the points you discussed. This gesture will help your case and pave the way for future meetings. Part of becoming an effective advocate for limited government is building good relationships with legislative staff members. If you have a history of good meetings, even though your legislator may not agree with you, you will likely have more access to your representatives or senators than the average person. This makes your voice, and that of your group, much louder and more influential.

ATTEND TOWN HALL MEETINGS

Elected officials often host town hall meetings in their districts to showcase their achievements and solicit feedback from their constituents. Such meetings are a prime opportunity for you to ask your lawmakers to state their position on an agenda of lower taxes, less government, and more freedom, on the record and in an open and public forum. Town hall meetings are held throughout the year, especially during congressional recesses. FreedomWorks has always encouraged our members to attend town hall meetings during the Presidents' Day Recess, Easter Recess, Memorial Day Recess, Independence Day Recess, August Recess, and so on. The town hall forums became much more popular in August 2009 when thousands of Americans attended town halls and effectively slowed down the Obamacare bill by providing an overwhelming opposition to the proposals before Congress. One of the investigators of the town hall protests of 2009 was Connecticut activist Bob MacGuffie, who wrote a now infamous memo called "Rocking the Town Halls." Bob's memo was picked up by leftist media outlets and caused many leftists in politics and the media to go apoplectic, accusing regular Americans of being "thugs," "un-American," and "evil-mongers." Bob's memo, however, did not preach thuggery but rather a very good tactical guide for taxpayers to make their voices heard.

Here are some things to remember about town hall meetings:

GET ON THE INVITATION LIST TO ATTEND THE MEETINGS: Bring as many members of your group who can attend. Write your lawmakers and explain that you are a local activist leader. Ask to be put on the invitation list for their town meetings and ask to bring members of your chapter. If they do not have such a list, ask for information on the next meeting. When you receive word that a town hall meeting is scheduled, be sure to make plans to attend, and share this information with the members of your chapter through e-mail, Facebook, group meetings, or other forms of communication.

Often, especially after August 2009, some cowardly congresspeople and senators will not make their town hall meetings public or will not do town halls at all. This is an opportunity for your group to wage a brief public relations campaign to convince the legislator to hold open forums for his or her constituents. We encourage groups to write letters to the editors of local newspapers and call local talk-radio stations, asking why the legislator refuses to either hold town halls or refuses to make the details public. One thing FreedomWorks activists have done in the past is to organize a town hall meeting of their own and then invite the legislator to come. If he or she refuses or ignores the invitation, the group should have a table, empty chair, microphone, and name plate for the legislator. Then invite the press to attend the town hall meeting and fill the room with a hundred constituents. Let the press report the headline, CONGRESSMAN X A NO-SHOW FOR TOWN HALL MEETING, CONSTITUENTS OUTRAGED. Then you can turn up the heat and follow up the news stories with more letters to the editor from other constituents who are disappointed that the legislator is not even listening to what the people have to say.

PREPARE QUESTIONS AHEAD OF TIME: If your legislator does hold town hall meetings, be sure to prepare ahead of time. Have specific questions in mind. Ask for your legislator's position on a specific bill or issue that you care about. Make sure to get an answer from him or her. And use your phone or video camera to record the answer so that you can post it on YouTube, Facebook, and your blog. There is no better way to hold politicians accountable for statements they make than to record them and post them online.

GET AN ANSWER: Make sure you are the first one to the microphone and that your group members are close behind. Often the first few questions in the town hall will define the entire event. Ask your question clearly and directly and expect a direct answer. If your legislator sidesteps or doesn't answer your question, calmly repeat it. For example, "Do you support fundamental tax reform or maintaining the status quo?" Be prepared for spin but always have some other folks ready to ask the same question a different way, in order to get a real answer on an issue. Your group should also applaud if the congressman gives an answer that they like, or shake their heads and say no if the congressman says something that they disagree with. Be polite but firm. Be respectful, but don't be afraid to be animated and passionate.

FOLLOW UP WITH A LETTER: Whether you had the opportunity to ask your question or not, follow up with a letter to your legislator. Let him or her know you attended their last town hall meeting. Ask your question in your letter if you didn't have an opportunity to do so at the meeting. This letter will ensure your lawmakers take you and your views seriously, and will allow for you to obtain a written response addressing your concerns.

One final note about town hall meetings: Freedom-Works always encourages a multipronged effort when it comes to these now more high-profile events. Have one team of activists stand outside the town hall meeting and hold up signs with questions or statements that make your points. This will help to set the tone of the town hall meeting inside and will give the media something to report on that includes your messages. It will make a powerful narrative if the people attending the town hall and the media both see protesters outside and concerned citizens inside the meeting with similar messages.

Limited government activists did this well during many of the town hall meetings across the country, and this model should be replicated even more as we continue to battle against big government in America.

7. RECRUITING

However you decide to spend most of your recruiting efforts, the key is to just do it. The important thing is that you steadily increase your membership numbers until you reach critical mass and even beyond that. If your community has a thousand people in it, and a certain percentage seem to lean toward fiscal conservatism, set an ambitious, multiyear goal to sign up whatever percentage that is. There are lots of opportunities to recruit new members to your cause, some of them mundane and others fun. It is hard work, but it will pay dividends when all of your efforts are that much bigger and better because you have a true enemy of limited government activists with you on local, state, and national battles.

ACTIVIST SPOTLIGHT: DIANA REIHMER

When the housing market crashed, so did the value of the Diana Reihmer's Philadelphia home. Unable to sell and unwilling to risk retirement on their quickly devaluing assets, the Reihmer's dreams of taking it easy were put on hold. "That's the way it is," Diana said. "You make your plans and life happens. We cut our budget and I went back to work. That was my bailout plan."

Meanwhile, politicians in Washington were bailing out companies and rewarding people who had made bad investments in homes and mortgage-backed securities. The contrast between her experience and the soft landings for Citigroup, GM, and AIG were galling.

Diana decided she needed to get involved. "My main motivation was to educate my fellow citizens on our constitutional roots. If a society wants to remain free, they need to hold their elected officials accountable and we as a nation were not doing a good job of that."

Diana started a Facebook page and connected with other concerned Philadelphians. She soon applied for a permit and planned a Tea Party. "More than five hundred people stood in the wind and the rain, just down the road from the birthplace of the constitution," Diana recalled. The Philadelphia tea party was off and running.

"I'm often asked what I mean when I say we need to take our country back," Diana said. "My answer is no to a time, but to a commitment to constitutional principles. Take it back to an ethos of entrepreneurship, self-reliance, and community."

As Diana reflects back on the 9/12 Taxpayer March on Washington, she speaks with emotion. "We met the most amazing people. Heading to Washington, I connected with new friends from across the country. As we arrived in Washington, someone thanked me for helping her find her voice—that this movement has allowed her to get out and express herself. It was overwhelming."

For a couple of good books that will help you understand recruiting people and marketing your group and its events, check out *Dedication and Leadership* by Douglas Hyde and *The Conservative Revolution: How to Win the Battle for College Campuses* by Brendan Steinhauser. When recruiting new members, be sure to get their name, e-mail address, cell phone number, and home address. The more information you have, the easier it will be to keep in touch with them and figure out how best to mobilize them in your community. Build a database to store all of your group contact information, and invest in a service that allows you to send mass e-mails. Also, make sure that FreedomWorks has their information so that we can stay directly in touch with them on all of our local, state, and national campaigns.

EVENTS

Events are a great way to tell people about your group and recruit them. Holding small and simple events or participating in another organization's event are excellent opportunities to recruit additional members. There are a lot of great event ideas out there, but here are just a few: movie night, book clubs, book signings, teach-ins, cocktail parties, general meetings, special issue meetings, candlelight vigils, protests, guest speakers, conferences, conventions, grassroots leadership training seminars, virtual events, and so on.

Fund-raisers such as bake sales or potluck dinners are excellent recruitment events that will not only help you raise money for your chapter but will also attract new people who will want to get involved in your group. Also, spreading the word about your chapter's rallies, precinct walks, and lobbying trips are great ways to keep your chapter active and attract new volunteers. An organization must be active to keep the interest of its members and to show its value in belonging. Some organizations remain stagnant after elections, and this is a big mistake. Your group should be active every week of the year, with a different focus as you go along. You don't want to keep doing the exact same things, so change it up, be open to new ideas, and add some fun and creativity to the mix.

Participating in another organization's event can take on many forms. It may involve being on a panel, attending a meeting or conference, or providing information at a booth. Always remember to take recruitment materials with you wherever you go. In addition to information about your group and your issues, you also should have sign-up sheets, and be sure someone is there to answer questions. Conferences are a great opportunity to network with other activists from across the country, share ideas, see your favorite speakers, and get political training from experts. You should encourage your group to attend when it can, and if there isn't

a state convention of some kind already, consider starting one. FreedomWorks is working with its coalition
allies in the movement to organize conventions of like-
minded groups in various states. These conventions are
open to members of the political parties, but the focus
is on policy and ideas, not parties. If there isn't already
a convention of this kind in your state, get in touch with
us and we will help advise you on putting one together.

GROUP MEETINGS

While the main reason for holding group meetings is
to keep the group informed and engaged, they also
have value as a recruitment device. As group meetings become regular community "happenings" that are
open to the public, more people will begin attending.
Try to run short, efficient meetings, and stick to your
agenda. Offer food and beverages, hand out literature
for educational purposes, and focus on welcoming new
attendees. Consider having guest speakers from time to
time, and be sure to introduce your group members to
all the resources the liberty movement has to offer.

LOCAL ISSUE BATTLES

Local issue battles can often act as the building blocks
of recruitment. Many FreedomWorks activists got

their start with us by fighting a local tax hike or fighting for spending controls at the local level. Our North Carolina chapter has done amazing work fighting for property rights. In spring 2009, Roy Loflin and Kathy Hartkopf launched the Orange Candy Tax Revolt with only a few yard signs and a small meeting. Within weeks more than 1,000 people were showing up to learn how to challenge tax hikes, and the revolt quickly swept to other counties around the state. These local issue battles are often the most important, have a better chance of success, and motivate the most people to join your group and take action. Local battles can do wonders for recruitment purposes, in addition to winning policy battles. Utilize grassroots petitions, both online and hardcopy, to sign up as many people as possible who share your values and sentiments. Then, you will have a big, local army ready to mobilize at the next city council meeting or county commissioners meeting.

We started relying heavily on sending out e-mail alerts to our contacts about legislation and local politics. For anyone doing this, I recommend keeping the e-mails as professional, fact-filled, and concise as possible. Also, be sure to send them out as sparingly as you can afford so that contacts will not be turned off and consider you a spammer. We had two special elections during the spring and early summer of 2009. Through e-mail we informed our contacts of ways to meet candidates and get involved in the political process. We sent out candidate questionnaires and shared the answers with our contacts and posted them on our Web site.

—CHRISTIE CARDEN, HUNTSVILLE, ALABAMA

THE INTERNET

Similar to the FreedomWorks Web site, your chapter's Web site can be an excellent recruitment tool. Be sure to keep the content on your site fresh and current. Also, make sure the contact information for your chapter (such as e-mail address, street address, and phone number) is prominently displayed and accurate. The more interesting and informative your Web site is, the more traffic you will get. This will help you develop an online interest in your chapter and find new volunteers. Be sure to take advantage of the boom in online social media and start a local Facebook group or Ning network. You can also organize on a state level by using teaparty.freedomworks.org. Twitter allows you to post information that is seen by everyone following you. If you build a big following on twitter.com, you will be able to share information with a huge number of people. Many people use a smartphone nowadays and have Twitter and Facebook applications. This is the wave of the future, and all liberty-loving activists should employ these new tools to organize, market their events, and spread the message of liberty.

MEMBERSHIP BROCHURES

For us, this is the FreedomWorks trifold membership brochure and a postage-paid envelope. It is Freedom-Works' stand-alone recruitment piece. Our volunteer chapter leaders receive copies, and if someone wants to find out more about FreedomWorks, this is what we give to them. The membership brochure provides quick and concise explanations of what FreedomWorks is doing and why we are effective. By getting someone to sign up using the insert, you will accomplish not only the recruitment goal discussed in this section, but you will be well on your way to achieving your fund-raising goals.

PERSONAL CONTACT

Of all the recruitment tactics, FreedomWorks recognizes the personal touch as the most important. When you are selling your local group, what you do, and your effectiveness, don't forget that you play a huge role in whether the person you approach ultimately decides to get involved. Someone is much more likely to join your group and become an active member after having a conversation with you. You might meet potential recruits at the grocery store, your kids' school, church, the post office, a community center, your university, or any other number of places. You should see yourself as a revolutionary missionary for liberty, and see everyone you meet as a potential recruit for the cause of liberty.

What do you say when you mention your group and the person seems receptive to knowing more? You should attest to your group's effectiveness, explain personal experiences, and answer questions. This works much better than simply reading an informational brochure or telling someone to browse the Web site. These other tools are good leave-behinds that can be in more places than you can, but they are no substitute for personal contact. You can bet that of the volunteers you recruit, the best, hardest-working, and most valuable will be the ones you talk to in a personal, one-on-one manner.

FELLOW MEMBER RECRUITMENT

Your best volunteers will act as key recruiters who will help spread the word and lend credibility to your local group. Getting information out about your chapter through word of mouth is effective because of the afore-mentioned personal touch. Be sure to remind your top officers and chapter members that they have a critical role to play as recruiters for the chapter. It is advisable to set some ambitious goals and then hold one another accountable.

When it comes to recruiting people for your group, there is no simple, all-encompassing set of guidelines. The results all differ depending on individual circum-stances. But it is often helpful to see real examples and consider how it has worked in the past.

Group leaders should apply whatever methods seem to work the best in their location.

> *By concentrating on activism while not being too overbearing in our communication with the folks who have joined us over the last twelve months, we have grown our membership from two friends who were worried about the direction of our country to an or-ganization that now has enough leverage to partici-pate in national events, as well as make changes within the local GOP. This year, for the first time ever in our district, we are having an open primary, a congressional race where the people will once again have a direct say in who will represent them.*
> —ANA PUIG, BUCKS COUNTY, PENNSYLVANIA

8. RETAINING ACTIVISTS

Retaining the volunteers who will become actively involved with your group goes hand-in-hand with recruiting them. Signing people for FreedomWorks and getting them involved with your chapter is great, but if their involvement goes no further than their initial excitement over FreedomWorks, they are less of an asset to your chapter. Retaining volunteers from battle to battle will truly build and strengthen your chapter and FreedomWorks. This section outlines some of the best ways to retain your chapter's volunteers and keep them active in the fight for lower taxes, less government, and more freedom.

CALLS TO ACTION

What FreedomWorks calls a "call to action" is probably the most important component to retaining your chapter's volunteers. This is an action related to a FreedomWorks campaign that gives people something productive to do. It can take the form of a petition, a letter to an elected official, or calling people to get them to attend a rally.

Calls to action keep the members of your chapter engaged and retain them as FreedomWorks volunteers. National issues such as taxes and federal spending will have calls to action sent from the national

FreedomWorks office. However, if you feel your chapter needs more to do or there is a specific local issue that you want to work on, let us know. We can help you brainstorm and come up with a variety of useful activities for your chapter and provide the materials and policy expertise to make them successful.

Now everything we do is posted on our Facebook page as well as our Twitter feed. So you can keep informed on the latest call to action there as well. The key is to help us make our calls to action go viral, reaching millions of people in a very short time.

ACTIVIST SPOTLIGHT: RYAN HECKER

When the housing market collapsed, Ryan Hecker was living in Houston, Texas, with his wife and baby. A Harvard-trained lawyer, Hecker has always been a fiscal conservative. Like so many others, he had grown disillusioned with the Republican party in recent years.

"Where was the commitment to big ideas?" he said. "Reagan would not have been content to piddle around with tax credits and deductions. He inspired a new generation of young people to take the big issues head on, not tinker at the edges."

Ryan remembered reading about the Contract with America in 1994. In high school at the time, he thought it was big and bold. The contract showed commitment to good governance and restraint in government growth. But as the years passed, he realized it came from the top down and was doomed to eventual compromise by career politicians.

So he thought about what he could do to bring back good governance and economic conservatism. "What if we the citizens come up with a contract from America and gave it to the politicians?" he wondered. It was an exciting idea, but how could just one person get that idea out?

After his chance encounter at a rally on February 27, Ryan became a member of the Houston Tea Party Society. Persuaded that he was part of something bigger than he had ever imagined, he decided he was going to try and make a go of his idea for a "Contract from America." With a few thousand dollars out of his own pocket, Ryan began to build a Web site to collect ideas from activists. The ideas would be voted on and the top ten would form the contract. It would be a bottom-up, grassroots effort that would lay out the people's priorities. If a legislator wanted to be on the side of the people, they would have an opportunity to embrace a document created by the people.

With little fanfare, Ryan launched his idea on September 4, 2009, to be a place for the policy ideas of a movement.

FOLLOW UP, FOLLOW UP, FOLLOW UP

The more you meet with your members, call them, e-mail them, or write them, the more active they will be. For instance, you may meet someone who is interested in your group at a gathering in your community. Make sure to get their contact information and call them in a couple of days to see if they have any questions or if they want to help out with your chapter's latest event.

Any time you receive a contribution for your local FreedomWorks group, make sure you send it to the national office as soon as possible so that we can process it and send your donor a thank-you letter. In addition, FreedomWorks encourages personalized thank-you notes and calls from you. Similar to recruiting, personal contact is critical to retaining people, especially when it comes to donors.

ACTIVITIES AND EVENTS

Making sure your chapter always has something to do helps you to be more effective. It is also one of the best ways to retain your membership and keep them interested and active. While you will constantly be supported by the FreedomWorks national office, always try to focus on local issues as well; examples include

petition drives, precinct walks, rallies, and special lobbying visits to the state capital. You will have the greatest success if the members of your chapter view your
planned activities as important and a good use of their
time.

Planned events also help reinforce the importance
of your local group and retain the volunteers you have
recruited. Although they often take more time, planning, and expense, events get more attention and can be
fun. The FreedomWorks office can also make a policy
expert available to speak at one of your chapter meetings or other events. One of the most requested things
we do is teach grassroots leadership training seminars
around the country. These popular sessions will teach
your volunteers how to increase their effectiveness by
learning and applying the strategies and tactics of community organizing that have worked for centuries.

NATIONAL RETENTION PROGRAM

While nothing supersedes the importance of the personal relationships that you maintain, know that
FreedomWorks has a national retention program that
you can depend on to help you retain your chapter
members. This program includes products, activities,
and services offered by FreedomWorks to keep the volunteers in your chapter engaged.

Depending on the level of involvement of your chapter members, they will receive various materials and have the opportunity to participate in events throughout the year. Some of these include:

Reports and updates about what Freedom-Works is doing and what is going on in their state and in Washington, D.C.

FreedomWorks policy papers

Invitations to special FreedomWorks events

The opportunity to participate in strategy calls

A variety of different calls to action including petition drives, get-out-the-vote efforts, and contact-your-legislator campaigns

When it comes to retention, nothing can replace the volunteer group leader as the consistent, on-the-ground presence for your chapter. However, the national retention program should complement your efforts to keep your volunteers engaged and ready for the next fight.

9. FUND-RAISING

One of the hardest tasks that you will encounter as a group leader will be the role you play as a fund-raiser for your chapter. The tireless energy needed to fuel your grassroots efforts will get you far, but you will need money. Many of the various activities and events that your chapter will engage in require funding in addition to planning and hard work for them to be successful.

The national FreedomWorks office will assist you in covering costs where we can, but raising money specifically for your chapter will further enhance and expand its scope of activities and events. As a result, Freedom-Works offers the following tips to help get you started toward successfully raising funds for your chapter.

HAVE A FUND-RAISING GOAL

Working with the FreedomWorks national office, establish a realistic fund-raising goal and physically outline how you plan to get there in the form of a brief fund-raising plan. Periodically, discuss your plan during your conversations with the FreedomWorks' national office. FreedomWorks can help you figure out what is working with your plan and what needs to be adjusted.

CREATE LISTS

As you get started looking at who you think would be likely donors to your chapter, start to create lists. Begin with those people you know and create a list of friends, family, church members, business acquaintances, members of service clubs, and so on, who know you. Draw from sources such as Christmas card lists, Rolodexes, and address books. Upon doing this, create two categories, ideological- and issue-based, and classify your prospective donors as one or the other. Your ideological list will comprise the donors who will give because they understand FreedomWorks and our mission, what our long-term goals are, and why it is important for your chapter to succeed. Issue-based donors will most likely be businesses that have a stake in the issue that you are working on at the time. After you have created your lists, you can then begin to form a plan of attack.

MAIL OUTREACH

Fund-raising through the mail can be a great revenue source, but it is also very tricky. The components that make fund-raising through the mail successful are generally involved, expensive, and based on volume, so you will probably not be using this tactic as a primary source of fund-raising for your chapter. But you should still be on the lookout for issues that have wide-reaching impact and could make a good fund-raising letter. In some instances, the FreedomWorks national office could be able to put a budget together to help you mail on such an issue. In addition, lists that include donors to other organizations and/or campaigns are key to acquire in the event that a mailing opportunity arises.

ACTIVIST SPOTLIGHT: TOM GAITENS

To Tom Gaitens, the Constitution is more than just a document; it is the spirit of the nation. "I don't understand someone who spends years as an academic to become a 'constitutional scholar.' The document is four pages, just read the thing and you understand exactly what the founders laid out for this nation."

During the TARP debate, Tom was busy organizing demonstrations outside both Republican and Democrat offices. "The politicians were desperate to do something, even though what they were doing was the wrong thing. Without a second thought, they will spend a trillion dollars of other people's money if it gives them the political cover of 'doing something.' The immediate impact of the stimulus bill was to make Congress feel better about their reelection prospects. That's a hell of a price tag for making 435 congressmen feel better. I wonder if they are so careless with their own money."

Tom organized an activist training session in Tampa at the end of January 2009. He focused on how President Obama used grassroots tactics to win caucus states and beat Hillary Clinton's political machine. Tom talked about the importance of holding both Democrats and Republicans accountable at all levels of government. One of the people in the audience was Mary Rakovich.

"TARP was the spark and the stimulus package was the gasoline," Tom explained. "When Santelli went on his rant, we were ready. For people like me working to get activists involved in the process, it all came together. I feel like this is a second American Revolution. We're ready to rediscover the importance of personal and economic freedom and not a moment too late."

TELEPHONE CAMPAIGNS

This is the primary tool you will use to reach out to the people on the lists you have created. Depending on those you are calling and how well you know them, you might feel more comfortable at first just talking about your group, how you have taken on the responsibility of a group leader and created a chapter in the area, and what you are trying to accomplish. Then you can ask the prospective donor if he or she would like to set up a meeting to talk more about how they can help and get involved. During that meeting make your fund-raising pitch or ask the person to join your advisory committee. Make sure before your meeting that you make use of the fund-raising experts in FreedomWorks' national office. They can help you better prepare for the meeting and work out your "pitch" to the prospective donor.

DEVELOP AN ADVISORY COMMITTEE

In many of the states that FreedomWorks has been actively involved on the ground, the advisory committee approach to fund-raising has proven quite effective. An advisory committee consists of a group of people (including members of the business community, prominent individuals, and other active persons) within the community whose responsibility it is either to give themselves or to solicit gifts for your chapter, generally in person. Make a list of the individuals you feel would be good candidates for your chapter's advisory committee and the possible chairs and cochairs. When you approach prospective members, be sure to stress the importance of your chapter and its work, but also be specific about what you expect from them. You have a goal, and the members of your advisory committee should be committed to helping you reach it.

SPECIAL EVENTS

Included here are all breakfasts, lunches, dinners, bar-becues, wine and cheese receptions, birthday parties, house parties, raffles, and other functions that can help you raise money for your group. In your fund-raising plan think about the kinds of events that you want to have throughout the year, how many you want (or think you will need), and when you want to have them and then draw up a quick time line or calendar. This will discipline you into making sure that these events are held.

Remember that these events are different from others your chapter will engage in because the primary goal is fund-raising, not impacting policy or drawing a crowd. Keep your costs as low as possible; no higher than 20 percent of the amount you hope to raise. It is key to use your imagination and not your checkbook to make these events memorable; examples abound in our various state chapters of successful low-cost fund-raisers. In North Carolina, selling tickets to rocking chair raffles, hot dog dinners, and pancake breakfasts have all yielded positive results. In Alabama, a series of

From the stage at the Washington, D.C., Tax Day Tea Party on April 15, 2010. *Photo by Terry Kibbe*

barbecues raised money, captured media attention, and promoted the chapter within the state. In all cases, the events first attracted people and raised funds because they were centered around specific issues, but had additional value because participants got to know Freedom-Works and became familiar with the local group.

ONLINE FUND-RAISING

The wave of the future in fund-raising for grassroots groups is online donations. Online donations can be in the form of e-mail solicitations that direct people to our Web site, "money bombs" to raise a lot of money in the same day, a static donation button on our Web site, and fund-raising campaigns for a specific event or effort. FreedomWorks has had the most success by asking for various amounts of money that appeal to a broad cross section of society. We will ask for as little as $5 and as much as $5,000 online. The key is to convince potential supporters that we are fighting hard for the liberty movement and that their money will be well spent and not wasted.

If you plan to help us raise money online or raise money directly for your local group, the same rules and best practices apply. And utilizing online tools for fund-raising is one of the easiest, quickest ways to support your efforts. A group that does not take online donations is missing the boat and should immediately open up a PayPal account to start accepting funds.

We have a Web site and made an agreement with our web page designer and host to provide their services at no cost in exchange for our promotion of their companies. To date, their services are easily valued at over $5,000. We also have a similar agreement with a flag supplier, who donates a small portion of his sales to our organization. We do have a line of T-shirts, bumper stickers, and other assorted patriotic items that are available for purchase.

I have personally asked family, friends, and all in attendance at any of our events or speaking engagements to donate what they can. This has amounted to approximately $3,000. It is explained to them that we are willing and able to fight this fight, but it is the funding that enables us to do so. Some of our donations have come in denominations as small as $5 and some as much as a few hundred. We have even had a congressman make a donation to the cause. Most people cite the tight economy as their reason for not donating or for donating so little.

—GREG FETTIG, HOOSIER PATRIOTS

Acknowledgments

This book could not have been written without the hard work, perseverance, and patriotic spirit of untold thousands of citizens who make up the Tea Party movement. Each of them has contributed to the story line about this peaceful revolution against the political establishment that nourishes, and then profits from, big government. We would like to thank all of the local Tea Party leaders who took time to tell their stories to us and other members of the FreedomWorks staff. Thanks also go to the thousands of folks who responded to our "Why We Marched" e-mail request and submitted for us their personal stories and recollections just days after September 12, 2009. All of these accounts, collected, are the bits of decentralized knowledge that tell the real story about that historic day.

We would also like to thank the contributors of pictures, including Terry Kibbe, Michael Beck, and others, who provided photo documentation of the sheer size, wonderful diversity, and joy of the crowds who attended the Taxpayer March on Washington and other Tea Party events across the country.

Many thanks to William Morrow/HarperCollins: our editor, Peter Hubbard, and the entire team, including, Liate Stehlik, Lynn Grady, Dee Dee DeBartlo, Jean Marie Kelly, and Shawn Nicholls.

We also want to thank Jon Yarian and the Pinkston Group who guided our sometime hectic writing process, often working with us through the night to deliver the book on time.

Finally, we offer our appreciation and respect for all of our colleagues at FreedomWorks, who best reflect the values, commitment, ability, and old-fashioned work ethic that is the Tea Party ethos. This book could not have happened without the research and support of Max Pappas, Wayne Brough, Brendan Steinhauser, and Adam Brandon. Adam also served as the relentless agitator in chief for this project, and it would never have happened without his tireless—and hopelessly naive—advocacy for the book.

Notes

CHAPTER 1: THE SILENT MAJORITY SPEAKS

13 "I guess the rest is history": This and all quotes from Mary Rakovich are from a personal interview with Adam Brandon, April 15, 2010.

23 national debt stood at $5.6 trillion: http://www.treasury direct.gov/govt/reports/pd/histdebt/histdebt_histo5 .htm.

23 amount had nearly doubled to $10 trillion: http://www .treasurydirect.gov/NP/BPDLogin?application=np.

23 more than $18 trillion: "Budget of the U.S. Government Fiscal Year 2010," May 2009, page 3.

23 60 percent of Gross Domestic Product: http://www .usgovernmentspending.com/federal_debt_chart.html.

23 more than 100 percent by 2012: http://www.cbo.gov/ ftpdocs/100xx/doc10014/03-20-PresidentBudget.pdf.

23 France's 2010 debt-to-GDP ratio: http://blogs.wsj.com/ brussels/2010/04/23/sovereign-debt-uk-vs-france.

24 $1.9 trillion in 2009: http://www.treasurydirect.gov/ govt/reports/pd/histdebt/histdebt_histo5.htm.

24 $2.1 trillion in revenue: "Budget of the U.S. Government Fiscal Year 2010," May 2009.

26 . . . driving '54 Chevys: *Squawk Box*, February 19, 2009.

CHAPTER 2: THE AMERICAN REVOLUTIONARY MODEL

29 "It was now evening, and I immediately dressed . . .": James A. Hawkes, *Retrospect of the Boston Tea-Party with a Memoir of George R. T. Hewes* (New York: S. S. Bliss, printer, 1834), 38.

37 "The tea party concept has gained significant traction . . .": Brian Stelter, "Reporter Says His Outburst Was Spontaneous," *New York Times*, March 3, 2009.

38 "There was no official count, but the crowd spilled . . .": *St. Louis Post-Dispatch*, February 26, 2009.

42 Joseph Galloway threatened to derail momentum: Galloway's speech can be found in the *Journals of the Continental Congress*, vol. 1 (Washington, D.C.: Library of Congress), 44–48.

43 "crowds around Carpenters' Hall soon heard . . .": A. J. Langguth, *Patriots: The Men Who Started the American Revolution* (New York: Simon & Schuster, 1989), 213.

43 "He eats little . . .": Ibid.

44 "I shall have great advantage . . .": Fawn Brodie, *Thomas Jefferson: An Intimate History* (New York: Norton, 1998), 124.

46 "the people are the only sure reliance . . .": Ibid.
46 "The preservation of the sacred fire of liberty . . .": Lance Banning, *The Sacred Fire of Liberty: James Madison and the Founding of the Federal Republic* (Ithaca, N.Y.: Cornell University, 1995).
47 Winston Churchill would later say: Justin D. Lyons, "Winston Churchill's Constitutionalism: A Critique of Socialism in America," http://www.heritage.org/Research/Reports/2009/05/Winston-Churchills-Constitutionalism-A-Critique-of-Socialism-in-America.
48 "It does not require a majority to prevail . . .": Greory Olinyk, *Resonation: Enlightened Government for We the People* (Garden City, N.Y.: Morgan James Publishing, 2006), 22.

CHAPTER 3: BAILOUT

49 "The straw that broke the back . . .": Personal interview with Adam Brandon, April 19, 2010.
52 6 percent to 1 percent: Federal Reserve, Federal Funds Effective Rate, "H.15, Selected Interest Rates," available at https://www.federalreserve.gov/releases/H15/date.htm.
52 "Too many dollars were churned out . . .": Judy Shelton, "Loose Money and the Derivative Bubble," for *Wall Street Journal* Symposium: Did the Fed Cause the Housing Bubble? March 27, 2009, 13 (available at http://online.wsj.com/article/SB123811225716453243.html).
53 A respected classical liberal scholar: Ludwig von Mises, *The Theory of Money and Credit* (Indianapolis: Liberty

Fund Press 1912/1980) and *Human Action* (New Haven: Yale University Press, 1949).

53 "... such a boom is bound to collapse ...": Ludwig von Mises, *Omnipotent Government* (New Haven: Yale University Press, 1944), 251.

55 banks "have tons of money left ...": Brian M. Carney, "The Credit Crisis Is Going to Get Worse," *Wall Street Journal,* July 5, 2008, A9.

56 Indeed, a recent release by the U.S. Census Bureau: "Residential Vacancies and Ownership in the First Quarter 2010, U.S. Census Bureau, April 26, 2010, www.census.gov/hhes/www/housing/hvs/qtr110/files/ q110press.pdf.

56 "... these government-sponsored enterprises have exposed taxpayers ...": Martin Reiser, "CSE Joins HOMEEC Coalition to Bring Needed Oversight to Fannie Mae and Freddie Mac," March 9, 2000. CSE (Citizens for a Sound Economy) was a predecessor organization of FreedomWorks.

56 "... more people ... exaggerate a threat ...": House Financial Services Committee, *The Treasury Department's Views on the Regulation of Government Sponsored Enterprises, Hearing Before the Committee on Financial Services, U.S. House of Representatives,* 108th Cong., 1st sess., September 10, 2003, 3.

57 "Remember the S & L banking scandal ...": Freedom-Works, "Top Ten Welfare Queens, 2006," August 8, 2006, available at http://www.freedomworks.org/publi cations/top-ten-welfare-queens-2006.

59 "Corrections are not all bad": John A. Allison to Congress, September 30, 2008, available at http://www .freedomworks.org/uploads/allison.pdf.

60 "owes more to Benito Mussolini...": Gerald P. O'Driscoll Jr., "An Economy of Liars," *Wall Street Journal,* April 20, 2010.

61 "there are no atheists in foxholes...": Quoted in Peter Baker "A Professor and a Banker Bury Old Dogma on Markets," *New York Times,* September 21, 2008, A1.

61 "abandoned free market principles...": President George W. Bush, interview with CNN reporter Candy Crowley, December 16, 2008, http://www.cnn.com/2008/POLITICS/12/16/bush.crowley.interview/index.html#cnnSTCVideo.

63 "the Secretary is authorized to take such actions...": Emergency Economic Stabilization Act, Title I, Sec. 101 (c), http://www.govtrack.us/congress/billtext.xpd?bill=h110-1424.

64 "...we must meet until this crisis is resolved": Sen. John McCain, "America Faces an Historic Crisis," Real Clear Politics, September 24, 2008, http://www.realclearpolitics.com/articles/2008/09/its_time_to_come_together_to_s.html.

67 "Congress has given the Secretary far-reaching power...": FreedomWorks Foundation, "Constitutional Infirmities of the Emergency Economic Stabilization Act of 2008 (EESA), Issue Analysis no. 124, January 13, 2009, http://www.freedomworks.org/files/policyanalysis.pdf.

68 fathers of the separation of powers: See John Locke, *Two Treatises of Government,* edited by Peter Laslett (New York: New American Library, 1965), especially the Second Treatise, Sections 143, 144, 150, and 159. See also Charles Montesquieu, *The Spirit of Laws* (Cambridge: Cambridge University Press, 1989).

68 "accumulation of all powers . . . in the same hands . . .":
James Madison, "The Particular Structure of the New
Government and the Distribution of Power Among Its
Different Parts," *The Federalist 47*, January 30, 1788,
available at http://www.constitution.org/fed/federa47
.htm.

68 Indeed, the TARP Congressional . . . been severely critical: Elizabeth Warren, "Accountability for the Troubled
Asset Relief Program," Congressional Oversight Panel,
January 9, 2009, available at www.cop.senate.gov/
reports/library/report-010909.cop.cfim.

69 "While the purpose of [the TARP legislation] is to
stabilize . . .": Treasury Secretary Henry Paulson, "Statement on Stabilizing the Automotive Industry," HP-1332,
U.S. Department of the Treasury, December 19, 2008.

70 "*If* we possess all the relevant information . . .": F. A.
Hayek, "The Use of Knowledge in Society," in *Indi-
vidualism and Economic Order* (Chicago: University of
Chicago Press, 1948), 77.

72 "The constitutional questionability of some provisions
is worrying . . .": Stuart Butler and Edwin Meese, III,
"The Bailout Package: Vital and Acceptable," The Heritage Foundation, September 29, 2008.

75 ". . . inaction is not a good option": https://www.afpmo
.org/index.php?id=6475.

78 "grassroots reaction is visceral . . .": Lisa Lerer, "Bailout Critics Say They're Losing," Politico.com, September 24, 2008, http://www.politico.com/news/
stories/0908/13809.html.

81 "economic freedom means the freedom to succeed . . .":
Terence P. Jeffrey, "Pence Calls on Conservatives to

Oppose Bailout That 'Nationalizes' Bad Mortgages," CNSNews.com, September 28, 2008, http://www .cnsnews.com/news/article/36463.

81 "lawmakers say they have received hundreds of calls . . .": Sarah Lueck and Michael Phillips, "Lawmakers Weigh Political Risks of Stance on Bailout Plan." *Wall Street Journal*, September 26, 2008, A6.

83 "As a public choice professor, I used to begin class each semester . . .": Dick Armey, "My Vote: NO," National Review Online, September 29, 2008, available at http://article.nationalreview.com/372981/my-vote-no/ dick-armey.

84 "If ye love wealth better than liberty . . .": Samuel Adams, speech, August 1, 1776.

85 "Americans' anger is in full bloom . . .": Sheryl Gay Stolberg, "Constituents Make Their Bailout Views Known," *New York Times*, September 25, 2008, A27.

86 "Exempts from the excise tax . . .": Lawrence H. White, "The Financial Bailouts: 'See the Needle and the Damage Done,'" *The Freeman 59*, no. 2, March 2009, http:// www.thefreemanonline.org/featured/the-financial-bail outs-%E2%80%9Csee-the-needle-and-the-dam age-done%E2%80%9D/#.

CHAPTER 4: WHAT WE STAND FOR

91 The Austrian economist Joseph Schumpeter noted: Joseph A. Schumpeter, *Capitalism, Socialism, and Democracy* (New York: Harper & Brothers, 1942).

92 the "seen and the unseen": Frédéric Bastiat, "That Which Is Seen and That Which Is Not Seen," in *The Bastiat Collection*, 2 vols. (Auburn, Ala.: Ludwig von Mises Institute, 2007), 2:1–48.

95 "... What are their names?": Personal recollection of Dick Armey and Phil Gramm, confirmed with Gramm.

96 2009 budget deficit was 10 percent of GDP: Jackie Calmes, "$1.4 Trillion Deficit Complicates Stimulus Plans," *New York Times*, October 17, 2009, A1.

96 interest on the debt alone: Jeanne Sahadi, "$4.8 trillion: Interest on U.S. Debt," *CNN Money.com*, December 20, 2009, http://money.cnn.com/2009/11/19/news/econ omy/debt_interest/index.htm?cnn=yes.

96 He replied "a republic if you can hold it": Notes of Dr. James McHenry, *American Historical Review* 11 (1906), 618.

CHAPTER 5: THE STATUS QUO LASHES OUT AT THE TEA PARTY

101 "As veteran Russia reporters...": Mark Ames and Yasha Levine, "Backstabber: Is Rick Santelli High On Koch?" in Megan McArdle, "Playboy Dips a Toe into Investigative Journalism," http://www.theatlantic.com/ business/archive/2009/03/playboy-dips-a-toe-into-in vestigative-journalism/4770/.

102 FreedomWorks as the "Armey of Darkness": Paul Krugman, "Armey of Darkness," *New York Times*, April 11,

2009, http://krugman.blogs.nytimes.com/2009/04/11/armey-of-darkness.

102 "They're Astroturf (fake grass roots) events . . .": Paul Krugman, "Tea Parties Forever," *New York Times*, April 13, 2009, A21.

102 "This [Tea Party] initiative is funded by the high end . . .": KTVU interview with Nancy Pelosi, April 15, 2009, http://thinkprogress.org/2009/04/15/pelosi-astro turf.

104 "Dick Armey's group is out there . . .": Press gaggle by Press Secretary Robert Gibbs, http://www.presidency .ucsb.edu/ws/index.php?pid=86538.

105 ". . . things that you could not do before": Gerald Seib, "In Crisis, Opportunity for Obama," *Wall Street Journal*, November 21, 2008, A2.

106 "It is now evident," they argued, "that an ugly campaign . . .": Nancy Pelosi and Steny Hoyer, "'Un-American' Attacks Can't Derail Health Care Debate," *USA Today*, August 10, 2009.

106 "I'm not afraid of August": Press conference with Nancy Pelosi, July 23, 2009, http://www.politico.com/blogs/glennthrush/0709/Pelosi_Im_not_afraid_of_August .html.

107 CNN's Anderson Cooper opted for dirty jokes: "Cable Anchors, Guests Use Tea Parties as Platform for Frat House Humor," http://www.foxnews.com/politics/2009/04/16/cable-anchors-guests-use-tea-par ties-platform-frat-house-humor.

107 managed to say the words *teabag* and *teabagger*: "Rachel Maddow's Disgusting Homoerotic Tea Party 'News'

Report," http://www.ihatethemedia.com/rachel-mad dow-disgusting-homoerotic-tea-party-news-report.

108 "helped create the teabaggers...": Jonathan Alter, *The Promise: President Obama, Year One* (New York: Simon and Schuster, 2010).

108 "They're carrying swastikas and symbols like that...": Interview with Nancy Pelosi, www.youtube.com/ watch?v=pFSZiG14GOU.

109 "... our own right-wing domestic terrorists...": http:// corner.nationalreview.com/post/?q=YWY5ZmYwNTk wMjE0MTMxYTk2ZjQyOTZhMmI0ZWQ3MzY.

109–110 "... we value our liberties...": Peter Hamby, "Axel- rod Suggests 'Tea Party' Movement Is 'Unhealthy,'" April 19, 2009, http://politicalticker.blogs.cnn.com/2009/04/19/ axelrod-suggests-tea-party-movement-is-unhealthy/?fbid =JiSECAjMn8i.

110 "... that picture of him in the *Post* today...": Presi- dent Bill Clinton, "The Tragedy of Oklahoma City 15 Years Later," April 16, 2010, http://www.americanpro gressaction.org/events/2010/04/inf/clinton.pdf.

111 "This is about hating a black man...": Keith Olbermann interview with Janeane Garofalo, *Countdown*, April 16, 2009, http://www.washingtontimes.com/weblogs/ back-story/2009/apr/17/liberal-actress-says-tea-par ties-were-racist/.

111 "... an overwhelming portion of the intensely demon- strated animosity...": Brian Williams, interview with President Jimmy Carter, *NBC Nightly News*, September 15, 2009, http://www.cnn.com/2009/POLITICS/09/15/ carter.obama/index.html.

113 "The public is outraged about the president's policies . . .": "Statement of Project 21 Fellow Deneen Borelli on Allegations of Racism Against Critics of Obama Policies," September 16, 2009, http://www.nationalcenter.org/PR-Racism_Obama_091609.html.

114 "Personally, I'm not a fan of this movement": David Brooks, "The Tea Party Teens," *New York Times*, January 5, 2010, A21.

115 "There is an argument floating . . .": David Brooks, "No U-Turns," *New York Times*, March 29, 2007.

115 ". . . tactics of the New Left": David Brooks, "The Wal-Mart Hippies," *New York Times*, March 5, 2010, A27.

116 ". . . evil is introduced into society by corrupt elites and rotten authority structures": http://www.nytimes.com/2010/03/05/opinion/05brooks.html.

117 "People of the same trade seldom meet together . . .": Adam Smith, *The Wealth of Nations*, http://www.econlib.org/library/Smith/smWN4.html#I.10.82

118 "Normal, nonideological people are less concerned . . .": Brooks, "No U-Turns."

119 "Now I'm not a true Republican . . .": Amy Gardner, "Republicans in Utah Direct Anger at Former Party Favorite Bennett," *Washington Post*, May 1, 2010, http://www.washingtonpost.com/wp-dyn/content/article/2010/04/30/AR2010043000770.html.

120 "a proposal for time travel . . .": Michael Gerson, "Doers vs. Undoers," townhall.com, February 26, 2010, http://townhall.com/columnists/MichaelGerson/2010/02/26/doers_vs_undoers?page=full&comments=true.

121 ". . . conservatism has been reduced to sound bites": Steven F. Hayward, "Is Conservatism Brain-Dead?" *Washington Post,* October 4, 2009.

121 "opposition to Barack Obama and the Democratic Congress has sparked a resurgence . . .": Brink Lindsey, "Right Is Wrong," *Reason* magazine, August 2010.

122 these informal networks take advantage of what philosopher Michael Polanyi: *Personal Knowledge: Towards a Post-critical philosophy"* (Chicago: University of Chicago Press, 1958).

123 "It seems that Greenspan, Bernanke, Fannie, Freddie . . .": David Boaz, "What Caused Atlas Shrugged Sales to Soar?" http://www.cato-at-liberty .org/2009/05/18/what-caused-atlas-shrugged-sales-to-soar/.

129 As Saul Alinsky teaches, "change comes from power . . .": Saul Alinsky, *Rules for Radicals: A Practical Primer for Realistic Radicals* (New York: Random House, 1971; New York: Vintage Books, 1989), 113.

129 ". . . Democrats have vilified the 'tea party' . . .": Neil Munro, "The Tea Party: Do Not Boil," *National Journal,* April 24, 2010.

130 Hank Sheinkopf, a Democratic political consultant": Ibid.

131 views of the "Tea Party" (48 percent): "Tea Party 48% Obama 44%," Rasmussen Reports, http://www .rasmussenreports.com/public_content/politics/gen eral_politics/april_2010/tea_party_48_obama_44.

CHAPTER 6: ACHIEVING CRITICAL MASS

144 "We won't get a public option...": Fred Barbash, "Robert Reich Calls for March on Washington in Support of Public Option," Politico.com, August 18, 2009, http://www.politico.com/news/stories/0809/26224 .html.

145 Sophia Elena, a video blogger: http://www.youtube. com/watch?v=HXo-13UQKcw.

146 "... it had received a bomb threat": Teddy Davis, "There's a 'Big Ol T.E.A. Party' Coming to Washington," September 11, 2009, http://blogs.abcnews. com/thenote/2009/09/theres-a-big-ol-tea-party-com ing-to-washington-.html.

150 "Everyone was in great spirits...": E-mail submitted to FreedomWorks "Why I Marched" initiative.

153 "I loved seeing people of all ages marching . . .": E-mail submitted to FreedomWorks "Why I Marched" initiative.

154 250,000 strong: http://www.lib.lsu.edu/hum/mlk/srs216 .html.

154 "FreedomWorks held a Capitol Hill demonstration yesterday...": "Democratic Memo: 9/12 Project Might Draw Two Million People to D.C. Tomorrow," September 11, 2009, http://hotair.com/archives/2009/09/11/ democratic-memo-912-project-might-draw-two-mil lion-people-to-dc-tomorrow.

155 "I was stunned by the number of people . . .": E-mail submitted to FreedomWorks "Why I Marched" initiative.

155 "as many as one million people...": David Gardner, "A Million March to U.S. Capitol to Protest Against 'Obama the Socialist,'" *London Daily Mail*, September 14, 2009, http://www.dailymail.co.uk/news/worldnews/article-1213056/Up-million-march-US-Capitol-protest-Obamas-spending-tea-party-demonstration.html.

156 "...their dc demonstration attracted 30,000 people": http://tpmdc.talkingpointsmemo.com/2009/09/freedomworks-cuts-estimate-for-crowd-at-its-912-rally-by-one-half.php.

156 "Authorities in the District...": Emma Brown, James Hohmann, and Perry Bacon Jr., "Lashing Out at the Capitol," *Washington Post*, September 13, 2009.

157–158 "probably well more than 850,000 in the crowd": Charlie Martin, "March on Washington: How Big Was the Crowd?" *Pajamas Media*, September 14, 2009.

158 "We overheard people talking...": E-mail submitted to FreedomWorks "Why I Marched" initiative.

159 "I will be fifty-one years old...": E-mail submitted to FreedomWorks "Why I Marched" initiative.

161 "The huge, polite, focused crowd was like a fresh wind...": E-mail submitted to FreedomWorks "Why I Marched" initiative.

CHAPTER 7: WHY WE MUST TAKE OVER THE REPUBLICAN PARTY

162 "Both parties seem to be...": Janet Hook, "New York Race at Epicenter of a GOP Mutiny," *Los Angeles Times*, October 27, 2009.

167 "This is not a good year . . .": This and all other quotes regarding the poll are from "CNN Poll: Anti-incumbent Fever at Record High," CNN.com, February 24, 2010.

167 ". . . PUBLIC SERVANTS END-OF-JOB COUNSELING . . .": *Grandma's Not Shovel-Ready: Signs from 9/12 and the Tea Parties of 2009* (Wilmington, Del.: Let Freedom Ring Publishing, 2010).

169 ". . . channel this tidal shift in American public opinion . . .": Mark Tapscott, "Tea Party Leaders are Wasting Opportunity to Replace Congress," *Washingon Examiner*, September 3, 2009.

170 "sign up for your local political party . . .": Erick Erickson, "Put Down the Protest Signs and Pick Up the Campaign Signs," RedState.com, September 2, 2009, http://www.redstate.com/erick/2009/09/02/put-down-the-protest-signs-and-pick-up-the-campaign-signs/.

170 "Would you guys knock off an incumbent Republican . . .": Chris Matthews, interview with Matt Kibbe, *Hardball*, MSNBC December 10, 2009.

176 In 1912, the Progressive Party was formed: James Chace, *1912: Wilson, Roosevelt, Taft & Debs—The Election That Changed the Country* (New York: Simon & Schuster, 2004), 168.

176 thirteen U.S. House members in office: Amos R. E. Pinchot and Helene Maxwell Hooker, *History of the Progressive Party, 1912–1916* (New York: New York University Press, 1958).

178 "The Year of the Woman": http://emilyslist.org/who/history/index.html.

179 700,000 donating members: http://cdn.moveonpac.org/content/pdfs/onepageaboutus.pdf.

180 ". . . the face of power in America": Andrea Seabrook, "Powerful PAC: EMILY's List Turns 25," *Morning Edition*, NPR, April 30, 2010.

182 "companies that have been awarded taxpayers' money . . .": "TARP Recipients Paid Out $114 Million for Politicking Last Year," OpenSecrets.org, February 4, 2009, http://www.opensecrets.org/news/2009/02/tarp-recipients-paid-out-114-m.html.

183 "Even in the best economic times . . .": Ibid.

183 "The taxpayers are a vast group . . .": Mancur Olson, *The Logic of Collective Action: Public Goods and the Theory of Groups*, rev. ed. (Cambridge, Mass.: Harvard University Press, 1971/1965), 165–66.

187 "The small-government candidate in the Republican Party . . .": Adam Nagourney and Jeremy W. Peters, "GOP Moderate, Pressed by Right, Abandons Race," *New York Times*, November 1, 2009, A1.

188 "Dede is more liberal than the Democrat": Janet Hook, "New York Race at Epicenter of a GOP Mutiny," *Los Angeles Times*, October 27, 2009.

188 She was in third place: Daily Kos / Research 2000 NY-23 Poll, http://www.dailykos.com/statepoll/2009/10/28/NY/408.

188 blaming the "hate and lies and the deceitfulness": Michelle Breidenbach, "Dede Scozzafava Says 'Hate, Lies' Wore Her Down," Syracuse.com, http://www.syracuse.com/news/index.ssf/2009/11/dede_scozzafava_says_hate_lies.html.

190 ". . . I am writing to let you know . . .": http://watertowndailytimes.com.

192 "Ms. Scozzafava fit the model of candidate . . .": Nagourney and Peters, "GOP Moderate, Pressed by Right."

CHAPTER 8: THE NEW CENTER OF AMERICAN POLITICS

201 "I mean, when you have parents asking neighbors . . .": Matt Clemente, e-mail message to Adam Brandon, April 12, 2010.

202 "In 2010 all politics are national": Rachel Martin, interview with Rick Barber, *Good Morning America*, January 24, 2010, http://abcnews.go.com/video/playerIndex?id=9647848.

219 "Florida will be one of the clearest tests . . .": E. J. Dionne Jr., "A Florida Test for the GOP," *Washington Post*, May 14, 2009.

221 "an inspiring leader for the next generation . . .": "Former U.S. House Majority Leader Dick Armey Endorses Marco Rubio for U.S. Senate," July 14, 2009, http://www.marcorubio.com/former-u-s-house-majority-leader-dick-armey-endorses-marco-rubio-for-u-s-senate.

224 "It is a damn outrage": David Brooks, *Meet the Press*, May 9, 2010.

CHAPTER 9: WE ARE A MOVEMENT OF IDEAS, NOT LEADERS

226 "Pick the target, freeze it . . .": Saul Alinsky, *Rules for Radicals: A Practical Primer for Realistic Radicals* (New York: Random House, 1971; New York: Vintage Books, 1989), 130.

226 Hayek's weighty notion: "spontaneous order": F. A. Hayek, *The Constitution of Liberty* (1960; repr., Chicago: University of Chicago Press, 1978).

227 "Which of these systems . . .": F. A. Hayek, *The Use of Knowledge in Society in Individualism and Economic Order* (Chicago: University of Chicago Press, 1948, 1980), 79.

228 "I would rather be exposed to the inconveniences . . .": Thomas Jefferson to Archibald Stuart, 23 December 1791. The Thomas Jefferson Papers, Series 1: General Correspondence, 1651–1827, Library of Congress, Washington, D.C.

228 "We can't give people a choice . . .": Personal recollection of Dick Armey.

EPILOGUE: CHANGING THE CULTURE IN TACOMA, WASHINGTON

238 "I don't expect politicians to solve anybody's problems . . .": Jann Wenner interview with Bob Dylan, "The Modern Times of Bob Dylan: A Legend Comes to Grips with His Iconic Status," *Rolling Stone*, September 7, 2006.

242 "Our own selves": Ibid.

244 "No, these are my people, Americans": Brent Baker, "White NBC Reporter Confronts Black Man at Tea Party Rally: 'Have You Ever Felt Uncomfortable?'" NewsBusters.org, April 16, 2010.

HARPER LUXE

THE NEW LUXURY IN READING

We hope you enjoyed reading
our new, comfortable print size and found it
an experience you would like to repeat.

Well – you're in luck!

HarperLuxe offers the finest in fiction and
nonfiction books in this same larger print size and
paperback format. Light and easy to read, HarperLuxe
paperbacks are for book lovers who want to see
what they are reading without the strain.

For a full listing of titles and
new releases to come, please visit our website:

www.HarperLuxe.com